ELECTROMAGNETIC RETARDATION AND THEORY OF RELATIVITY

Oleg D. Jefimenko
WEST VIRGINIA UNIVERSITY

ELECTROMAGNETIC RETARDATION AND THEORY OF RELATIVITY

NEW CHAPTERS IN THE
CLASSICAL THEORY OF FIELDS

Electret Scientific Company
Star City

 Requests for permission or for further information should be addressed to the publisher, Electret Scientific Company, P.O. Box 4132, Star City, West Virginia 26504, USA.

Library of Congress Catalog Card Number: 97-60547

ISBN 0-917406-21-4

Printed in the United States of America

PREFACE

This book is a sequel to my *Electricity and Magnetism*, 2nd ed., (Electret Scientific, Star City, 1989) and *Causality, Electromagnetic Induction, and Gravitation*, (Electret Scientific, Star City, 1992). It is a result of a further exploration of the classical theory of fields in search of heretofore overlooked relations between physical quantities and heretofore overlooked applications of the theory. The book is divided into two parts. The first part, Chapters 1 to 5, presents the fundamentals of the theory of electromagnetic retardation with emphasis on recently discovered relations and recently developed mathematical techniques. The second part, Chapters 6 to 11, presents the fundamentals of the theory of relativity based entirely on the theory of electromagnetic retardation developed in the first part.

Electromagnetic retardation is as yet a fairly obscure concept, and therefore an explanation of what it is and why a book needs to be written about it is in order.

Electric and magnetic fields propagate with finite velocity. Therefore there always is a time delay before a change in electromagnetic conditions initiated at a point of space can produce an effect at any other point of space. This time delay is called electromagnetic retardation. Recent studies have shown that electromagnetic retardation is of overriding importance for the general electromagnetic theory and, by extension, for the entire classical theory of fields. We now know that electromagnetic retardation manifests itself in many different ways including, but not limited to, electromagnetic cause-and-effect relations, electromagnetic waves generated by oscillating electric charges and currents, electromagnetic fields and potentials of time-dependent charge and current distributions, electromagnetic fields of moving charge distributions, mechanical relations between time-dependent or moving charges and currents, dynamics of atomic systems, time relations in moving electromagnetic systems, and the visual appearance of moving bodies. Perhaps the most important recently discovered aspect of the now evolving theory of electromagnetic retardation is that this theory leads to, and duplicates, many electromagnetic relations that are customarily considered to constitute consequences of relativistic electrodynamics. In fact, it is now clear that there exists an intimate relation between the theory of electromagnetic retardation and the theory of relativity. Obviously then, the phenomenon of electromagnetic retardation and its theoretical representation must be thoroughly understood and investigated.

In contrast with the theory of electromagnetic retardation, the theory of relativity is fairly familiar. However, as far as its scientific essence is concerned, the theory of relativity means different things to different people. It is important therefore to give a clear definition of the expression "theory of relativity" as it is used in this book.

In this book, "theory of relativity" (or "relativity theory," or simply "relativity") is used as a collective term for the body of equations, methods, and techniques whereby physical quantities measured in one inertial frame of reference can be correlated with physical quantities measured in any other inertial frame of reference.

As already mentioned, there exists an intimate relation between the theory of electromagnetic retardation and the theory of relativity. On the basis of this relation, all the fundamental equations of the theory of relativity, including equations of relativistic electrodynamics and relativistic mechanics, are derived in Chapters 6 to 8 in a natural and direct way from equations of the theory of electromagnetic retardation without any postulates, conjectures, or hypotheses. As a result, Maxwellian electromagnetism, electromagnetic retardation, and the theory of relativity are united in this book into one simple, clear, and harmonious theory of electromagnetic phenomena and of mechanical interactions between moving bodies.

An important consequence of the theory of relativity developed in the above manner is the revelation of certain basic errors in the interpretation and use of Einstein's special relativity theory. The nature of these errors and the ways to avoid them are explained in Chapter 9.

One of the most controversial elements of Einstein's special relativity theory is his idea of universal kinematic time dilation, according to which the rate of all moving physical and biological "clocks" are uniformly dilated in consequence of nothing other than the relative motion of the clocks. As is shown in Chapter 10, moving elementary electromagnetic clocks indeed run slower than the same stationary clocks, but their slower rate is a consequence of dynamic interactions and depends on both the velocity and the construction of the clocks.

An extension of the theory of relativity, as it is developed in this book, leads to a covariant theory of gravitation analogous to relativistic electrodynamics. This extension is presented in Chapter 11, the concluding chapter of the book.

Although the book presents the results of original research, it is written in the style of a textbook and contains numerous illustrative examples demonstrating various applications of the theory developed in the book. Therefore it can be used not only for independent reading, but also as a supplementary textbook in courses on electromagnetic theory and on the theory of relativity.

I am pleased to acknowledge with gratitude a stimulating exchange of correspondence with P. Hillion, J. J. Smulsky, V. N. Strel'tsov, and W. E. V. Rosser on some aspects of the theory of relativity, and with M. A. Heald on the subject of electromagnetic retardation.

I am very grateful to S. W. Durland and D. K. Walker for carefully reading the manuscript and for their most useful suggestions and recommendations.

Special thanks are due to Yu. G. Kosarev who believes that retardation is a universal phenomenon that should be properly treated in a new branch of physics which he proposes to call "retardics." His comments are highly appreciated.

Finally, I am very grateful to my wife Valentina for proofreading the numerous versions of the manuscript and for otherwise helping me to make the book ready for publication.

Oleg D. Jefimenko
April 14, 1997

CONTENTS

3 RETARDED INTEGRALS FOR ELECTRIC AND MAGNETIC FIELDS AND POTENTIALS OF MOVING CHARGES

4 ELECTRIC AND MAGNETIC FIELDS AND POTENTIALS OF MOVING POINT AND LINE CHARGES

7 THE ESSENTIALS OF RELATIVISTIC ELECTRODYNAMICS

8 FROM RELATIVISTIC ELECTROMAGNETISM TO RELATIVISTIC MECHANICS

9 COMMON MISCONCEPTIONS ABOUT RELATIVITY THEORY

10 THE RATE OF MOVING CLOCKS

11 GRAVITATION AND COVARIANCE

I

RETARDATION

1

RETARDED INTEGRALS AND OPERATIONS WITH RETARDED QUANTITIES

The fundamental laws of electromagnetism are represented mathematically by Maxwell's electromagnetic equations. The general solution of these equations for electromagnetic fields in a vacuum is expressed in terms of "retarded" field integrals which constitute the basic mathematical element in the general theory of time-dependent electromagnetic phenomena. A thorough understanding of the properties and use of retarded integrals is therefore indispensable for formulation and application of the theory. In this chapter we shall acquaint ourselves with retarded integrals and with operations involving quantities and expressions appearing in these integrals.

1-1. Vector Wave Fields and Retarded Integrals[1]

The vector wave field is the field of a vector **V** which satisfies the *inhomogeneous wave equation* (also known as the *general wave equation*)

$$\nabla \times \nabla \times \mathbf{V} + \frac{1}{c^2}\frac{\partial^2 \mathbf{V}}{\partial t^2} = \mathbf{K}(x,y,z,t), \qquad \text{(1-1.1)}$$

where **K** is some vector function of space and time which, for simplicity, will be assumed here to be zero outside a finite region of space (this differential equation constitutes a mathematical expression for a wave-like disturbance that propagates in space with the speed c).

An important property of a vector wave field is that this field can be represented by the *retarded field integral* and *retarded potentials*, as explained in the following theorem.

The Wave Field Theorem. A vector field $\mathbf{V}$ satisfying Eq. (1-1.1) and vanishing at infinity can be represented by the retarded integral

$$\mathbf{V} = -\frac{1}{4\pi}\int \frac{[\nabla'(\nabla' \cdot \mathbf{V}) - \mathbf{K}]}{r} dV', \qquad (1\text{-}1.2)$$

where the brackets are the "retardation symbol," to be explained below, and r is the distance from the *source point* $P'(x', y', z')$ where the volume element of integration, dV', is located to the *field point* $P(x, y, z)$ where $\mathbf{V}$ is being determined; the primed operator ∇' operates on the source-point coordinates only. (*Note*: The integration in the above integral is over all space; except when noted otherwise, the integration in all integrals that follow is also over all space.)

The derivation of Eq. (1-1.2) is mostly of historical interest and will not be presented here.[2] In lieu of the derivation we shall show in Example 1-2.3 that Eq. (1-1.1) is satisfied by $\mathbf{V}$ given by Eq. (1-1.2).

Corollary I. A vector field $\mathbf{V}$ satisfying Eq. (1-1.1), vanishing at infinity, and having zero divergence outside a finite region of space can be represented by the retarded scalar potential φ and the retarded vector potential $\mathbf{A}$ as

$$\mathbf{V} = -\nabla\varphi + \nabla \times \mathbf{A}, \qquad (1\text{-}1.3)$$

with φ and $\mathbf{A}$ given by

$$\varphi = \frac{1}{4\pi}\int \frac{[\nabla' \cdot \mathbf{V} + K_1]}{r} dV' + \varphi_0 \tag{1-1.4}$$

and

$$\mathbf{A} = \frac{1}{4\pi}\int \frac{[\mathbf{K}_2]}{r} dV' + \mathbf{A}_0, \tag{1-1.5}$$

where K_1 and $\mathbf{K}_2$ are the ordinary potentials of the function $\mathbf{K}$ of Eq. (1-1.1) (so that $\mathbf{K} = -\nabla K_1 + \nabla \times \mathbf{K}_2$), both vanishing at infinity, and φ_0 and $\mathbf{A}_0$ are arbitrary constants.

Corollary II. A vector field $\mathbf{V}$ satisfying Eq. (1-1.1), vanishing at infinity, and having zero divergence outside a finite region of space can be represented by the retarded scalar potential φ and the retarded vector $\mathbf{W}$ as

$$\mathbf{V} = -\nabla\varphi + \mathbf{W}, \tag{1-1.6}$$

with

$$\varphi = \frac{1}{4\pi}\int \frac{[\nabla' \cdot \mathbf{V}]}{r} dV' + \varphi_0 \tag{1-1.7}$$

and

$$\mathbf{W} = \frac{1}{4\pi}\int \frac{[\mathbf{K}]}{r} dV' + \mathbf{W}_0, \tag{1-1.8}$$

where φ_0 and $\mathbf{W}_0$ are arbitrary constants. The proof of these corollaries is presented in Examples 1-2.1 and 1-2.2.

The retardation symbol [ ] indicates a special space and time dependence of the quantities to which it is applied and is defined by the identity

$$[f] \equiv f(x', y', z', t - r/c), \tag{1-1.9}$$

where t is the time for which the retarded integrals are evaluated. Thus the value of a function placed between the retardation symbol [ ] is *not* that which the function has at the time t for which the integrals are evaluated, but that which it *had* at some earlier time $t' = t - r/c$, or, as one says, the function is *retarded.*

The integrals of retarded quantities, or *retarded integrals,* are mathematical expressions reflecting the phenomenon of "final signal speed" - that is, the fact that a certain time r/c must elapse before the results of some event at the point x', y', z' can produce an effect at the point x, y, z separated from the point x', y', z' by a distance r.

Retarded integrals are closely associated with the principle of causality. According to this principle, all present phenomena are exclusively determined by past events. Therefore equations depicting causal relations between physical phenomena must, in general, be equations where a present-time quantity (the effect) relates to one or more quantities (causes) that existed at some previous time. As we shall presently see, in electromagnetic theory retarded integrals are "causal equations" expressing electric and magnetic fields and potentials in terms of their causative sources: the electric charge density ρ and the electric current density $\mathbf{J}$.[3]

1-2. Mathematical Operations with Retarded Quantities

Mathematical manipulations with retarded integrals frequently require applications of the operator ∇ to retarded quantities. When applying ∇ to such functions, one should take into account that they depend on space coordinates not only explicitly, but also implicitly through

$$r = \{(x-x')^2+(y-y')^2+(z-z')^2\}^{1/2} \tag{1-2.1}$$

appearing in the retarded time $t' - r/c$. One also should take into account that ∇ may operate with respect to x, y, z coordinates as well as with respect to x', y', z' coordinates. Finally, one should take into account that a ∇ operation may be performed upon a retarded quantity taken at the instant t = constant as well as at the instant $t' = t - r/c$ = constant (the latter operation is identical with the corresponding operation upon the same *unretarded* quantity, combined with the *subsequent* "retardation" of the resulting expression by replacing in this expression t by $t - r/c$).

Let us designate an unspecified scalar or vector function $f(x', y', z', t)$, together with an appropriate multiplication sign, if needed, by X. To avoid ambiguities with ∇ operations involving X, we shall employ special notations, as follows. If an operation is to be performed with respect to primed coordinates, we shall use the primed operator ∇' in writing this operation, and we shall use the ordinary operator ∇ for designating operations with respect to unprimed coordinates. If an operation upon a retarded X is to be performed considering the retarded time $t - r/c$ as constant, we shall denote the operation as $[\nabla X]$ or $[\nabla' X]$, placing both the operator and the function upon which it operates between the retardation brackets, and we shall use the ordinary notations $\nabla[X]$ or $\nabla'[X]$ for operations upon retarded functions when these operations are to be performed considering the present time t, rather than $t - r/c$, as constant.

We shall frequently use expressions and operations involving the radius vector connecting a volume element dV' of an electric charge or current (the source point x', y', z') with the point of observation (the field point x, y, z). If this radius vector is directed toward the field point, we shall designate it as $\mathbf{r}$, if it is directed toward the source point, we shall designate it as $\mathbf{r}'$. Likewise, we shall designate the corresponding unit vectors as $\mathbf{r}_u$ and $\mathbf{r}'_u$. Observe that since $\mathbf{r} = (x - x')\mathbf{i} + (y - y')\mathbf{j} + (z - z')\mathbf{k}$ and $\mathbf{r}' = (x' - x)\mathbf{i} + (y' - y)\mathbf{j} + (z' - z)\mathbf{k}$, the vector $\mathbf{r}' = -\mathbf{r}$, so that the result of any operation upon $\mathbf{r}'$ or r' with ∇ or ∇' is the negative of the result of the same operation upon $\mathbf{r}$ or r, and the result of any operation upon $\mathbf{r}$, $\mathbf{r}'$, r or r' with ∇ is the negative of the result of the same operation with ∇'.

We shall now derive several useful operational equations for retarded functions. Let us consider the operation $\partial[\mathrm{X}]/\partial x' \mid_{y', z', t}$, where [X] is some retarded scalar or vector function.[4] Taking into account that retarded functions depend on x', y', and z' not only directly, but also indirectly through r, we can write

$$\frac{\partial[\mathrm{X}]}{\partial x'}\bigg|_{y',z',t} = \frac{\partial[\mathrm{X}]}{\partial x'}\bigg|_{y',z',t-r/c} + \frac{\partial[\mathrm{X}]}{\partial(t-r/c)}\bigg|_{x',y',z'} \cdot \frac{\partial(t-r/c)}{\partial x'}. \tag{1-2.2}$$

We can simplify the last expression by noting that

$$\frac{\partial[\mathrm{X}]}{\partial(t-r/c)}\bigg|_{x',y',z'} = \left[\frac{\partial \mathrm{X}}{\partial t}\right]\bigg|_{x',y',z'}, \tag{1-2.3}$$

and that, by Eq. (1-2.1),

$$\frac{\partial(t-r/c)}{\partial x'} = \frac{x-x'}{cr} = \frac{\cos\alpha}{c}, \tag{1-2.4}$$

where $\cos\alpha$ is the direction cosine of vector **r** with respect to the x axis (Fig. 1.1). We then obtain

$$\frac{\partial[\mathrm{X}]}{\partial x'}\bigg|_{y',z',t} = \frac{\partial[\mathrm{X}]}{\partial x'}\bigg|_{y',z',t-r/c} + \frac{\cos\alpha}{c}\left[\frac{\partial \mathrm{X}}{\partial t}\right]\bigg|_{x',y',z'}. \tag{1-2.5}$$

Analogous expressions can be obtained also for $\partial[\mathrm{X}]/\partial y' \mid_{x', z', t}$ and for $\partial[\mathrm{X}]/\partial z' \mid_{x', y', t}$. If we now multiply these expressions by the unit vectors **i**, **j**, and **k**, respectively, and then add them together, we obtain the following operational equation

$$\nabla'[\mathrm{X}] = [\nabla'\mathrm{X}] + \frac{\mathbf{r}_u}{c}\left[\frac{\partial \mathrm{X}}{\partial t}\right], \tag{1-2.6}$$

where

$$\begin{aligned}\mathbf{r}_u &= \frac{\mathbf{r}}{r} = \frac{\mathbf{i}(x-x') + \mathbf{j}(y-y') + \mathbf{k}(z-z')}{r} \\ &= \mathbf{i}\cos\alpha + \mathbf{j}\cos\beta + \mathbf{k}\cos\gamma\end{aligned} \tag{1-2.7}$$

is the unit vector directed along r toward the point x, y, z ($\cos\beta$

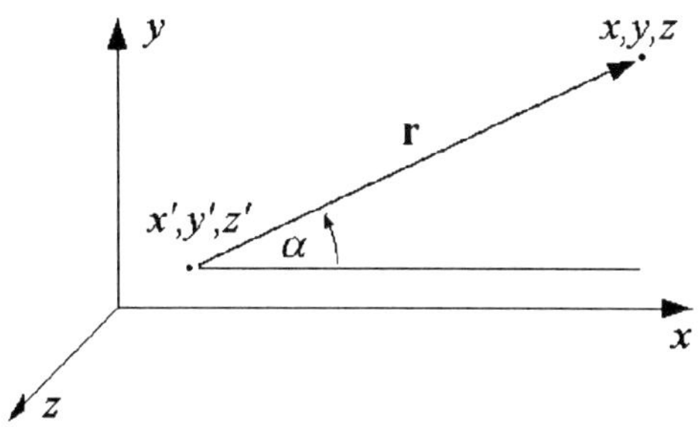

Fig. 1.1 The direction cosine of **r** *with respect to the x axis is* $\cos\alpha = (x - x')/r$.

and $\cos\gamma$ are the direction cosines of **r** with respect to the y and z axis, respectively).

In a similar manner we can obtain the corresponding equation for the unprimed ∇ (assuming that X does not explicitly depend on x, y, z)

$$\nabla[\mathrm{X}] = -\frac{\mathbf{r}_u}{c}\left[\frac{\partial \mathrm{X}}{\partial t}\right]. \tag{1-2.8}$$

Combining Eqs. (1-2.8) and (1-2.6), we obtain an equation correlating one unprimed ∇ operation with two primed ∇ operations

$$[\nabla'\mathrm{X}] = \nabla[\mathrm{X}] + \nabla'[\mathrm{X}]. \tag{1-2.9}$$

Differentiating $\nabla\{[\mathrm{X}]/r\}$ and using Eq. (1-2.9), we obtain the correlation

$$\nabla\frac{[\mathrm{X}]}{r} = -\frac{\mathbf{r}_u[\mathrm{X}]}{r^2} + \frac{\nabla[\mathrm{X}]}{r} = \frac{\mathbf{r}'_u[\mathrm{X}]}{r^2} + \frac{[\nabla'\mathrm{X}]}{r} - \frac{\nabla'[\mathrm{X}]}{r}, \tag{1-2.10}$$

and, combining the first and the last term of the last part of Eq. (1-2.10), we obtain a useful equation

$$\frac{[\nabla'\mathrm{X}]}{r} = \nabla\frac{[\mathrm{X}]}{r} + \nabla'\frac{[\mathrm{X}]}{r}. \tag{1-2.11}$$

Another useful equation is obtained by eliminating $\nabla[X]$ from the middle part of Eq. (1-2.10) by means of Eq. (1-2.8):

$$\nabla\frac{[X]}{r} = -\frac{\mathbf{r}_u[X]}{r^2} - \frac{\mathbf{r}_u}{rc}\left[\frac{\partial X}{\partial t}\right]. \tag{1-2.12}$$

Finally we note that, since

$$\frac{\partial[X]}{\partial(t-r/c)} = \frac{\partial[X]}{\partial t}, \tag{1-2.13}$$

we have, by Eqs. (1-2.3) and (1-2.13),

$$\left[\frac{\partial X}{\partial t}\right] = \frac{\partial[X]}{\partial t}. \tag{1-2.14}$$

▼

Example 1-2.1 Prove Corollary I to the wave field theorem, assuming that $\nabla \cdot \mathbf{V}$, K_1, and $\mathbf{K}_2$ are zero outside a finite region of space.

Expressing in Eq. (1-1.2) $\mathbf{K}$ as $\mathbf{K} = -\nabla K_1 + \nabla \times \mathbf{K}_2$ and using Eq. (1-2.11), we have

$$\begin{aligned}\mathbf{V} &= -\frac{1}{4\pi}\int\frac{[\nabla'(\nabla'\cdot\mathbf{V})-\mathbf{K}]}{r}dV' \\ &= -\frac{1}{4\pi}\int\frac{[\nabla'(\nabla'\cdot\mathbf{V})+\nabla'K_1-\nabla'\times\mathbf{K}_2]}{r}dV' \\ &= -\frac{1}{4\pi}\int\nabla\frac{[\nabla'\cdot\mathbf{V}+K_1]}{r}dV' - \frac{1}{4\pi}\int\nabla'\frac{[\nabla'\cdot\mathbf{V}+K_1]}{r}dV' \\ &\quad + \frac{1}{4\pi}\int\nabla\times\frac{[\mathbf{K}_2]}{r}dV' + \frac{1}{4\pi}\int\nabla'\times\frac{[\mathbf{K}_2]}{r}dV'.\end{aligned} \tag{1-2.15}$$

The second and the fourth integrals of the last expression can be transformed into surface integrals by using vector identities (V-20) and (V-21) (see Appendix for a list of vector identities). But since, by supposition $\nabla \cdot \mathbf{V}$, K_1, and $\mathbf{K}_2$ are zero outside a finite region of space, while the surface integrals are taken over all space, the integrals vanish. We thus have

$$\mathbf{V} = -\frac{1}{4\pi}\int \nabla \frac{[\nabla' \cdot \mathbf{V} + K_1]}{r} dV' + \frac{1}{4\pi}\int \nabla \times \frac{[\mathbf{K}_2]}{r} dV'. \quad (1\text{-}2.16)$$

Factoring ∇ out from under the integral signs (we can do so because the integration is with respect to primed coordinates, while ∇ operates upon the unprimed coordinates) and designating the resulting integrals as $\varphi - \varphi_0$ and $\mathbf{A} - \mathbf{A}_0$, we obtain Corollary I to the wave field theorem.

Example 1-2.2 Prove Corollary II to the wave field theorem.

As in the preceding example, we have

$$\begin{aligned} \mathbf{V} &= -\frac{1}{4\pi}\int \frac{[\nabla'(\nabla' \cdot \mathbf{V}) - \mathbf{K}]}{r} dV' = -\frac{1}{4\pi}\int \nabla \frac{[\nabla' \cdot \mathbf{V}]}{r} dV' \\ &- \frac{1}{4\pi}\int \nabla' \frac{[\nabla' \cdot \mathbf{V}]}{r} dV' + \frac{1}{4\pi}\int \frac{[\mathbf{K}]}{r} dV'. \end{aligned} \quad (1\text{-}2.17)$$

The second integral of the last expression is, as in Example 1-2.1, zero. We thus have

$$\begin{aligned} \mathbf{V} &= -\frac{1}{4\pi}\int \nabla \frac{[\nabla' \cdot \mathbf{V}]}{r} dV' + \frac{1}{4\pi}\int \frac{[\mathbf{K}]}{r} dV' \\ &= -\nabla\left(\frac{1}{4\pi}\int \frac{[\nabla' \cdot \mathbf{V}]}{r} dV'\right) + \frac{1}{4\pi}\int \frac{[\mathbf{K}]}{r} dV'. \end{aligned} \quad (1\text{-}2.18)$$

Designating the first integral as $\varphi - \varphi_0$ and the second integral as $\mathbf{W} - \mathbf{W}_0$, we obtain Corollary II to the wave field theorem.

Example 1-2.3 Show that $\mathbf{V}$ given by Eq. (1-1.2) satisfies Eq. (1-1.1)

Using vector identity (V-16), we can rewrite Eq. (1-1.1) as

$$\nabla^2 \mathbf{V} - \frac{1}{c^2}\frac{\partial^2 \mathbf{V}}{\partial t^2} = \nabla(\nabla \cdot \mathbf{V}) - \mathbf{K} = \mathbf{Z}, \quad (1\text{-}2.19)$$

where we have denoted $\nabla(\nabla \cdot \mathbf{V}) - \mathbf{K}$ as $\mathbf{Z}$ for simplicity.

Let us now divide the volume of integration in Eq. (1-1.2) into two parts: Vol_1 and Vol_2. Let Vol_1 be a very small region close to the point of observation, so that within this region the retardation can be neglected. We then have from Eq. (1-1.2)

$$\mathbf{V}_1 = -\frac{1}{4\pi}\int_{Vol\,1}\frac{\mathbf{Z}}{r}dV', \tag{1-2.20}$$

where the integral is not retarded. But this integral represents the well-known solution of the Poisson equation[5]

$$\nabla^2\mathbf{V}_1 = \mathbf{Z}. \tag{1-2.21}$$

The contribution of Vol_1 to $\nabla^2\mathbf{V}$ in Eq. (1-2.19) is therefore given by Eq. (1-2.21).

Let us now determine the contribution of Vol_2 to $\nabla^2\mathbf{V}$ in Eq. (1-2.19). From Eq. (1-1.2) we have

$$\nabla^2\mathbf{V}_2 = \nabla^2\left(-\frac{1}{4\pi}\int_{Vol2}\frac{[\mathbf{Z}]}{r}dV'\right) = -\frac{1}{4\pi}\int_{Vol2}\nabla^2\frac{[\mathbf{Z}]}{r}dV', \tag{1-2.22}$$

where we have placed ∇^2 under the integral sign, because ∇^2 operates upon the unprimed coordinates, while the integration is with respect to primed coordinates.

We can evaluate the last integral in Eq. (1-2.22) by integrating, in turn, the x, y, and z components of the integrand. Taking into account that ∇^2 can be expressed as $\nabla \cdot \nabla$, using Eqs. (1-2.12), (1-2.8), and (1-2.14), and remembering that $\nabla \cdot \mathbf{r} = 3$, $\nabla(1/r^n) = -(n/r^{n+1})\mathbf{r}_u$, and $\mathbf{r} \cdot \mathbf{r}_u = r$, we find, after somewhat lengthy but very simple calculations[6]

$$\nabla^2 V_{x2} = -\frac{1}{4\pi}\int_{Vol2}\frac{\partial^2[Z_x]}{rc^2\partial t^2}dV'. \tag{1-2.23}$$

Since similar equation can be obtained also for the y and z components of $\mathbf{V}_2$, Eq. (1-2.22) becomes

$$\nabla^2 \mathbf{V}_2 = -\frac{1}{4\pi}\int_{Vol2} \frac{\partial^2 [\mathbf{Z}]}{rc^2 \partial t^2} dV'. \tag{1-2.24}$$

Factoring out $\partial^2/c^2\partial t^2$, we have

$$\nabla^2 \mathbf{V}_2 = \frac{\partial^2}{c^2\partial t^2}\left(-\frac{1}{4\pi}\int_{Vol2} \frac{[\mathbf{Z}]}{r} dV'\right), \tag{1-2.25}$$

or, by Eq. (1-1.2), remembering that $\mathbf{Z} = \nabla(\nabla \cdot \mathbf{V}) - \mathbf{K}$,

$$\nabla^2 \mathbf{V}_2 = \frac{\partial^2 \mathbf{V}_2}{c^2 \partial t^2}. \tag{1-2.26}$$

The contribution of Vol_2 to $\nabla^2\mathbf{V}$ in Eq. (1-2.19) is therefore given by Eq. (1-2.26).

Adding now Eqs. (1-2.21) and (1-2.26), we obtain

$$\nabla^2(\mathbf{V}_1 + \mathbf{V}_2) = \frac{\partial^2 \mathbf{V}_2}{c^2 \partial t^2} + \mathbf{Z}. \tag{1-2.27}$$

Since Vol_1 can be made as small as we please compared to Vol_2, $\partial^2\mathbf{V}_1/c^2\partial t^2$ can likewise be made as small as we please compared to $\partial^2\mathbf{V}_2/c^2\partial t^2$. Therefore, assuming that $Vol_1 \ll Vol_2$, we can add $\partial^2\mathbf{V}_1/c^2\partial t^2$ to the right side of Eq. (1-2.27) without affecting the equation. We then have

$$\nabla^2(\mathbf{V}_1 + \mathbf{V}_2) = \frac{\partial^2 \mathbf{V}_2}{c^2 \partial t^2} + \frac{\partial^2 \mathbf{V}_1}{c^2 \partial t^2} + \mathbf{Z} = \frac{\partial^2}{c^2 \partial t^2}(\mathbf{V}_1 + \mathbf{V}_2) + \mathbf{Z}, \tag{1-2.28}$$

or

$$\nabla^2(\mathbf{V}_1 + \mathbf{V}_2) - \frac{\partial^2}{c^2 \partial t^2}(\mathbf{V}_1 + \mathbf{V}_2) = \mathbf{Z}, \tag{1-2.29}$$

so that $\mathbf{V}_1 + \mathbf{V}_2$, and therefore $\mathbf{V}$ given by Eq. (1-1.2) does indeed satisfy Eq. (1-1.1).

▲

References and Remarks for Chapter 1

1. This section closely parallels a similar section in the author's *Electricity and Magnetism*, 2nd ed., (Electret Scientific, Star City, 1989) pp. 46-52.

2. The solution of the scalar counterpart of Eq. (1-1.1) was first published by G. Kirchhoff in Ann. der Phys. und Chemie, **18**, p. 663 (1883). See also O. W. Richardson, *The Electron Theory of Matter* (Cambridge University Press, London, 1914) pp. 189-193 and R. B. McQuistan, *Scalar and Vector Fields* (Wiley, New York, 1965) pp. 292-305.

3. Causal relations in the domain of electromagnetic phenomena are analyzed in the author's book *Causality, Electromagnetic Induction, and Gravitation* (Electret Scientific, Star City, 1992).

4. The notation $\partial[X]/\partial x' \,|_{\,y',z',t}$ means "y', z', t are held constant."

5. The integral and the associated Poisson's equation are best known in connection with the magnetic vector potential produced by an electric current. See Ref. 1, pp. 363-364.

6. An alternative method of evaluating the Laplacian of the integral in Eq. (1-1.2) is to use spherical coordinates. See, for example, R. Becker and F. Sauter, *Electromagnetic Fields and Interactions* (Blaisdell, New York, 1964) pp. 280-281 or M. A. Heald and J. B. Marion, *Classical Electromagnetic Radiation*, 3rd ed., (Saunders, Fort Worth, 1995) pp. 258-260. Observe, however, that this method is based on a presumed spherical symmetry of the integrand in Eq. (1-1.2), and is therefore of limited validity.

2

RETARDED INTEGRALS FOR ELECTROMAGNETIC FIELDS AND POTENTIALS

A basic problem in electromagnetic theory is the obtaining of equations expressing electric and magnetic fields and potentials in terms of their causative sources: electric charges and currents. In the case of time-dependent systems, the most general equations expressing electric and magnetic fields and potentials in terms of charges and currents involve retarded integrals. Electric and magnetic fields and potentials expressed in terms of retarded integrals are called retarded electric and magnetic fields and potentials. In this chapter we shall derive several types of equations for retarded fields and potentials of time-dependent charge and current distributions and shall give examples of the use of these equations.

2-1. Maxwell's Equations and the Wave Field Theorem

The basic electromagnetic field laws are represented by four Maxwell's equations which, in their differential form, are[1]

$$\nabla \cdot \mathbf{D} = \rho \tag{2-1.1}$$

$$\nabla \cdot \mathbf{B} = 0 \tag{2-1.2}$$

$$\nabla \times \mathbf{E} = -\frac{\partial \mathbf{B}}{\partial t} \tag{2-1.3}$$

and

$$\nabla \times \mathbf{H} = \mathbf{J} + \frac{\partial \mathbf{D}}{\partial t}, \tag{2-1.4}$$

where **E** is the electric field vector, **D** is the electric displacement vector, **H** is the magnetic field vector, **B** is the magnetic flux density vector, **J** is the electric current density vector, and ρ is the electric charge density. For fields in a vacuum (the only fields with which we shall be concerned in this book), Maxwell's equations are supplemented by the two constitutive equations

$$\mathbf{D} = \varepsilon_0 \mathbf{E} \tag{2-1.5}$$

and

$$\mathbf{B} = \mu_0 \mathbf{H}, \tag{2-1.6}$$

where ε_0 is the permittivity of space and μ_0 is the permeability of space. (The names and designations of electromagnetic quantities used in this book are the same as those used in Ref. 1.)

In Maxwell's equations electric and magnetic fields are linked together in an intricate manner, and neither field is explicitly represented in terms of its sources. However, with the help of the vector wave field theorem introduced in Section 1.1 we can express each field in terms of its causative sources. To do so, we shall first convert Eqs. (2-1.1) - (2-1.4) into two inhomogeneous wave equations, thereby separating the two fields one from the other.

Taking the curl of Eq. (2-1.3) and using Eq. (2-1.6), we have

$$\nabla \times \nabla \times \mathbf{E} = -\frac{\partial}{\partial t}\nabla \times \mathbf{B} = -\mu_0 \frac{\partial}{\partial t}\nabla \times \mathbf{H}. \tag{2-1.7}$$

Eliminating $\nabla \times \mathbf{H}$ by means of Eq. (2-1.4) and using Eq. (2-1.5), we obtain

$$\nabla \times \nabla \times \mathbf{E} = -\mu_0 \frac{\partial \mathbf{J}}{\partial t} - \mu_0 \frac{\partial^2 \mathbf{D}}{\partial t^2} = -\mu_0 \frac{\partial \mathbf{J}}{\partial t} - \varepsilon_0 \mu_0 \frac{\partial^2 \mathbf{E}}{\partial t^2}. \quad (2\text{-}1.8)$$

Rearranging terms and replacing $\varepsilon_0\mu_0$ by $1/c^2$, we finally obtain

$$\nabla \times \nabla \times \mathbf{E} + \frac{1}{c^2}\frac{\partial^2 \mathbf{E}}{\partial t^2} = -\mu_0 \frac{\partial \mathbf{J}}{\partial t}. \quad (2\text{-}1.9)$$

Taking now the curl of Eq. (2-1.4) and using Eq. (2-1.5), we have

$$\nabla \times \nabla \times \mathbf{H} = \nabla \times \mathbf{J} + \frac{\partial}{\partial t}\nabla \times \mathbf{D} = \nabla \times \mathbf{J} + \varepsilon_0 \frac{\partial}{\partial t}\nabla \times \mathbf{E}. \quad (2\text{-}1.10)$$

Eliminating $\nabla \times \mathbf{E}$ by means of Eq. (2-1.3) and using Eq. (2-1.6), we obtain

$$\nabla \times \nabla \times \mathbf{H} = \nabla \times \mathbf{J} - \varepsilon_0 \frac{\partial^2 \mathbf{B}}{\partial t^2} = \nabla \times \mathbf{J} - \varepsilon_0 \mu_0 \frac{\partial^2 \mathbf{H}}{\partial t^2}. \quad (2\text{-}1.11)$$

Rearranging terms and replacing $\varepsilon_0\mu_0$ by $1/c^2$, we finally obtain

$$\nabla \times \nabla \times \mathbf{H} + \frac{1}{c^2}\frac{\partial^2 \mathbf{H}}{\partial t^2} = \nabla \times \mathbf{J}. \quad (2\text{-}1.12)$$

Equations (2-1.9) and (2-1.12) are the general electromagnetic wave equations for the electric and magnetic fields, respectively. Applying Eq. (1-1.2) (the vector wave field theorem) to Eqs. (2-1.9) and (2-1.12), we can write for the electric field

$$\mathbf{E} = -\frac{1}{4\pi}\int \frac{\left[\nabla'(\nabla' \cdot \mathbf{E}) + \mu_0 \frac{\partial \mathbf{J}}{\partial t}\right]}{r} dV', \quad (2\text{-}1.13)$$

and for the magnetic field

$$\mathbf{H} = -\frac{1}{4\pi}\int\frac{\left[\nabla'(\nabla'\cdot\mathbf{H})-\nabla'\times\mathbf{J}\right]}{r}dV', \qquad (2\text{-}1.14)$$

where **E** and **H** are determined for the instant t, and the quantities in the brackets are taken at the corresponding retarded time $t' = t - r/c$ (c is the velocity of light in a vacuum).

2-2. Solution of Maxwell's Equations in Terms of Retarded Integrals

According to Eqs. (2-1.1) and (2-1.5), $\nabla \cdot \mathbf{E} = \rho/\varepsilon_0$, and according to Eqs. (2-1.2) and (2-1.6), $\nabla \cdot \mathbf{H} = 0$. Applying these relations to Eqs (2-1.13) and (2-1.14) and noting that $\varepsilon_0\mu_0 = 1/c^2$, we obtain

$$\mathbf{E} = -\frac{1}{4\pi\varepsilon_0}\int\frac{\left[\nabla'\rho + \dfrac{1}{c^2}\dfrac{\partial\mathbf{J}}{\partial t}\right]}{r}dV' \qquad (2\text{-}2.1)$$

and

$$\mathbf{H} = \frac{1}{4\pi}\int\frac{[\nabla'\times\mathbf{J}]}{r}dV'. \qquad (2\text{-}2.2)$$

Equations (2-2.1) and (2-2.2) constitute solutions of Maxwell's equations for fields in a vacuum and represent the electric and magnetic fields in terms of their causative sources: the electric charge and current distributions.[2] Since the fields in Eqs. (2-2.1) and (2-2.2) are expressed in terms of retarded integrals, these fields are called *retarded fields*.

There are several special forms into which Eqs. (2-2.1) and (2-2.2) can be transformed. One such special form is obtained from Eqs. (2-2.1) and (2-2.2) by eliminating from them the spatial derivatives. This can be done as follows.

Writing Eq. (2-2.1) in terms of two integrals and using vector identity (V-33) to transform the first integral, we have

$$\begin{aligned}\mathbf{E} &= -\frac{1}{4\pi\varepsilon_0}\left\{\int \frac{[\nabla'\rho]}{r}dV' + \frac{1}{c^2}\int \frac{1}{r}\left[\frac{\partial \mathbf{J}}{\partial t}\right]dV'\right\} \\ &= -\frac{1}{4\pi\varepsilon_0}\left\{\int \nabla\frac{[\rho]}{r}dV' + \int \nabla'\frac{[\rho]}{r}dV' + \frac{1}{c^2}\int \frac{1}{r}\left[\frac{\partial \mathbf{J}}{\partial t}\right]dV'\right\}.\end{aligned} \tag{2-2.3}$$

The second integral in the last expression can be transformed into a surface integral by means of vector identity (V-20). But this integral vanishes, because ρ is confined to a finite region of space, while the surface of integration is at infinity. Transforming the integrand in the first integral by means of vector identity (V-34) and using $\mathbf{r}_u = \mathbf{r}/r$, we then obtain for the electric field

$$\mathbf{E} = \frac{1}{4\pi\varepsilon_0}\int\left\{\frac{[\rho]}{r^3} + \frac{1}{r^2c}\left[\frac{\partial\rho}{\partial t}\right]\right\}\mathbf{r}\,dV' - \frac{1}{4\pi\varepsilon_0 c^2}\int\frac{1}{r}\left[\frac{\partial\mathbf{J}}{\partial t}\right]dV'. \tag{2-2.4}$$

Similarly, applying vector identities (V-33) and (V-21) to Eq. (2-2.2), taking into account that there are no currents at infinity, and using vector identity (V-34), we obtain for the magnetic field

$$\mathbf{H} = \frac{1}{4\pi}\int\left\{\frac{[\mathbf{J}]}{r^3} + \frac{1}{r^2c}\left[\frac{\partial\mathbf{J}}{\partial t}\right]\right\}\times\mathbf{r}\,dV'. \tag{2-2.5}$$

Observe that in Eqs. (2-2.4) and (2-2.5) the vector **r** is directed toward the point of observation (the field point).

Equation (2-2.4) represents a generalization of the electrostatic Coulomb's field integral to time-dependent systems and reduces to that integral in the case of time-independent fields in a vacuum. Likewise, Eq. (2-2.5) represents a generalization of the Biot-Savart's integral for magnetic fields and reduces to that integral in the case of time-independent systems.[3]

Another form of the field equation for **E** can be obtained as follows. According to the conservation of electric charge law (the continuity law),[4]

$$\frac{\partial \rho}{\partial t} = -\nabla \cdot \mathbf{J}. \tag{2-2.6}$$

Therefore the contribution that $\partial\rho/\partial t$ makes to the first integral in Eq. (2-2.4) can be expressed as

$$\int \frac{1}{r^2 c}\left[\frac{\partial \rho}{\partial t}\right]\mathbf{r}\,dV' = -\int \frac{[\nabla' \cdot \mathbf{J}]}{r^2 c}\mathbf{r}\,dV'. \tag{2-2.7}$$

Using now vector identity (V-30) with $\mathbf{r}_u = \mathbf{r}/r$ for transforming the last integral, and using vector identity (V-8), we obtain

$$\begin{aligned}\int \frac{[\nabla' \cdot \mathbf{J}]}{r^2 c}\mathbf{r}\,dV' &= \int\left(\frac{\nabla' \cdot [\mathbf{J}]}{r^2 c}\mathbf{r} - \frac{\mathbf{r}\cdot[\partial \mathbf{J}/\partial t]}{r^3 c^2}\mathbf{r}\right)dV' \\ &= \int\left(\frac{\mathbf{r}}{c}\nabla' \cdot \frac{[\mathbf{J}]}{r^2} - \frac{\mathbf{r}}{c}[\mathbf{J}]\cdot\nabla'\frac{1}{r^2} - \frac{\mathbf{r}\cdot[\partial \mathbf{J}/\partial t]}{r^3 c^2}\mathbf{r}\right)dV'.\end{aligned} \tag{2-2.8}$$

Next, using vector identity (V-23), we transform the first term in the integrand of the last integral, obtaining

$$\int \frac{\mathbf{r}}{c}\nabla' \cdot \frac{[\mathbf{J}]}{r^2}dV' = \oint \frac{\mathbf{r}}{c}\left(\frac{[\mathbf{J}]}{r^2}\cdot d\mathbf{S}'\right) - \int\left(\frac{[\mathbf{J}]}{r^2}\cdot\nabla'\right)\frac{\mathbf{r}}{c}dV'. \tag{2-2.9}$$

Since the integration is over all space, and since there is no current at infinity, the surface integral in Eq. (2-2.9) vanishes. Applying vector identity (V-4) to the integrand of the remaining integral in Eq. (2-2.9) and remembering that a ∇' operation upon $\mathbf{r}$ is the negative of the same ∇ operation (see Chapter 1, p. 7), we then have

$$\int \frac{\mathbf{r}}{c}\nabla' \cdot \frac{[\mathbf{J}]}{r^2}dV' = \int \frac{[\mathbf{J}]}{cr^2}dV'. \tag{2-2.10}$$

From Eqs. (2-2.7), (2-2.8), (2-2.9), and (2-2.10), we obtain therefore

$$\int \frac{1}{r^2c}\left[\frac{\partial \rho}{\partial t}\right]dV' = -\int\left(\frac{[\mathbf{J}]}{r^2c} - \frac{\mathbf{r}}{c}[\mathbf{J}]\cdot\boldsymbol{\nabla}'\frac{1}{r^2} - \frac{\mathbf{r}\cdot[\partial\mathbf{J}/\partial t]}{r^3c^2}\mathbf{r}\right)dV'. \quad (2\text{-}2.11)$$

Substituting Eq. (2-2.11) into Eq. (2-2.4) and taking into account that $\boldsymbol{\nabla}'(1/r^2) = 2\mathbf{r}/r^4$, we finally obtain[5]

$$\mathbf{E} = \frac{1}{4\pi\varepsilon_0}\int\frac{[\rho]}{r^3}\mathbf{r}\,dV' \quad (2\text{-}2.12)$$
$$- \frac{1}{4\pi\varepsilon_0 c}\int\left\{\frac{[\mathbf{J}]}{r^2} - 2\mathbf{r}\frac{[\mathbf{J}]\cdot\mathbf{r}}{r^4} - \frac{\mathbf{r}}{r^3c}\left[\frac{\partial\mathbf{J}}{\partial t}\right]\cdot\mathbf{r} + \frac{1}{rc}\left[\frac{\partial\mathbf{J}}{\partial t}\right]\right\}dV'.$$

It is important to note that although in Eqs. (2-2.1)-(2-2.12) the charge density, the current density, and their derivatives are retarded, retardation can frequently be neglected, in which case the above equations can be used with ordinary (unretarded) charge density, current density, and their derivatives. Let us define the "characteristic time" of an electromagnetic system as the time T during which the charge density, the current density, or their temporal derivatives experience a significant change. For example, in the case of periodic charge and current variations, T may be assumed to be the period of the oscillations, and in the case of monotonously changing charges and currents, T may be assumed to be the time during which the charge density, the current density, or their temporal derivatives change by a factor of two. Let us now assume that the largest linear dimensions of the system under consideration is L. If T and L satisfy the relation

$$T \gg L/c, \quad (2\text{-}2.13)$$

then no significant change occurs in the system during the time that the electric or magnetic field signal moves across the system, and therefore the retardation in the propagation of the electric or magnetic fields within the system is negligible. In Section 2.5 we shall discuss in some detail electromagnetic effects in systems to which Eq. (2-2.13) applies.

▼

Example 2-2.1 A thin circular ring of radius a and cross-sectional area s carries a uniformly distributed charge q. At $t = 0$ the ring starts to rotate with constant angular acceleration α about its symmetry axis which is also the x axis of rectangular coordinates (Fig. 2.1). Find the electric and magnetic fields at a point x on the axis for $t > 0$.

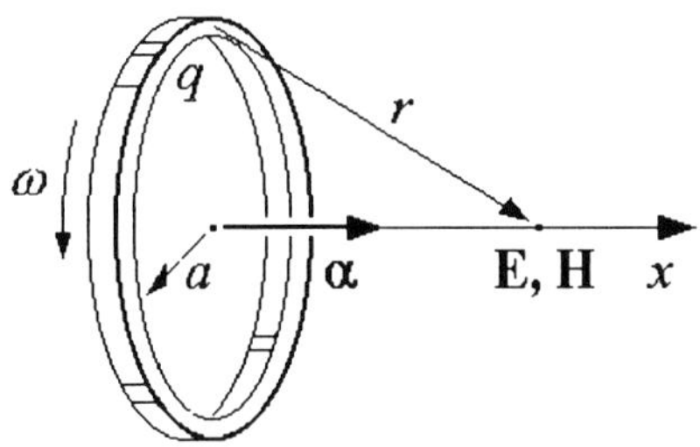

Fig. 2.1 Calculation of the electric and magnetic fields on the axis of a charged ring rotating with angular acceleration α.

The current density **J** created by the rotating ring is $\mathbf{J} = \rho\mathbf{v} = \rho\omega a\boldsymbol{\theta}_u = \rho\alpha t a\boldsymbol{\theta}_u$, where ρ is the charge density in the ring, ω is the angular velocity of the ring, and $\boldsymbol{\theta}_u$ is a unit vector in the circular direction (right-handed with respect to x). The time derivative of **J** is $\partial\mathbf{J}/\partial t = \rho\alpha a\boldsymbol{\theta}_u$. In terms of q, the current density and the derivative are $\mathbf{J} = (q\alpha t/2\pi s)\boldsymbol{\theta}_u$ and $\partial\mathbf{J}/\partial t = (q\alpha/2\pi s)\boldsymbol{\theta}_u$.

To find the electric field, we can use Eq. (2-2.4). Since $\partial\mathbf{J}/\partial t$ is in the circular direction, and since r is the same for all points of the ring, the second integral in Eq. (2-2.4) makes no contribution to the electric field on the axis (the contributions of any two volume elements on the opposite ends of a diameter cancel each other). Since the charge density does not depend on time, the contribution of the first integral is

$$\mathbf{E} = \frac{1}{4\pi\varepsilon_0}\int\frac{\rho}{r^3}\mathbf{r}\,dV', \tag{2-2.14}$$

which is identical with the expression for the electrostatic field produced by a stationary charge density ρ. The solution of Eq. (2-

2.14) for a charged ring is well known,[6] and therefore we shall reproduce it here without calculations. It is

$$\mathbf{E} = \frac{qx}{4\pi\varepsilon_0(a^2 + x^2)^{3/2}}\mathbf{i}. \qquad (2\text{-}2.15)$$

To find the magnetic field, we can use Eq. (2-2.5). Expressing [**J**] and [$\partial\mathbf{J}/\partial t$] in Eq. (2-2.5) in terms of q, α, s, and $\boldsymbol{\theta}_u$, we have

$$\begin{aligned}\mathbf{H} &= \frac{1}{4\pi}\int\left\{\frac{q\alpha(t - r/c)}{2\pi s r^3}\boldsymbol{\theta}_u + \frac{q\alpha}{r^2 c 2\pi s}\boldsymbol{\theta}_u\right\}\times \mathbf{r}\,dV' \\ &= \frac{1}{4\pi}\int\left\{\frac{q\alpha t}{2\pi s r^3}\boldsymbol{\theta}_u - \frac{q\alpha}{r^2 c 2\pi s}\boldsymbol{\theta}_u + \frac{q\alpha}{r^2 c 2\pi s}\boldsymbol{\theta}_u\right\}\times \mathbf{r}\,dV' \qquad (2\text{-}2.16) \\ &= \frac{1}{4\pi}\int\left\{\frac{q\alpha t}{2\pi s r^3}\boldsymbol{\theta}_u\right\}\times \mathbf{r}\,dV'.\end{aligned}$$

The current formed by the ring is filamentary. Its magnitude is $I = Js = q\alpha t/2\pi$. Since the current is filamentary, the volume element dV' in Eq. (2-2.16) can be written as sdl', where dl' is a length element along the circumference of the ring. Furthermore, we can combine $\boldsymbol{\theta}_u$ and dl' into the vector $d\mathbf{l}' = dl'\boldsymbol{\theta}_u$. We then have from Eq. (2-2.16)

$$\mathbf{H} = -\frac{1}{4\pi}\oint\frac{I}{r^3}\mathbf{r}\times d\mathbf{l}', \qquad (2\text{-}2.17)$$

which is identical with the expression for the magnetic field produced by a time-independent filamentary current I. The solution of Eq. (2-2.17) for a ring current is well known.[7] It is

$$\mathbf{H} = \frac{Ia^2}{2(a^2 + x^2)^{3/2}}\mathbf{i}, \qquad (2\text{-}2.18)$$

or, substituting $I = q\alpha t/2\pi$,

$$\mathbf{H} = \frac{q\alpha t a^2}{4\pi(a^2 + x^2)^{3/2}}\mathbf{i}. \qquad (2\text{-}2.19)$$

The surprising result of this example is that neither the electric nor the magnetic field on the axis of the rotating ring is affected by retardation.

Example 2-2.2 Electromagnetic waves can be generated by a radiating "electric dipole antenna." It consists of a piece of straight open wire which carries a current

$$I = I_0 \sin \omega t . \tag{2-2.20}$$

The current in the wire is produced by cutting the wire in the middle and connecting the two parts to a source of alternating current. If the length l of the antenna is much smaller than the wavelength of the generated waves, $l \ll \lambda = 2\pi c/\omega$, the antenna is called a "Hertzian dipole." In a Hertzian dipole the current is the same along the entire length of the antenna. Find the magnetic and electric fields produced by the Hertzian dipole shown in Fig. 2.2, at a large distance $r \gg l$ from the dipole.

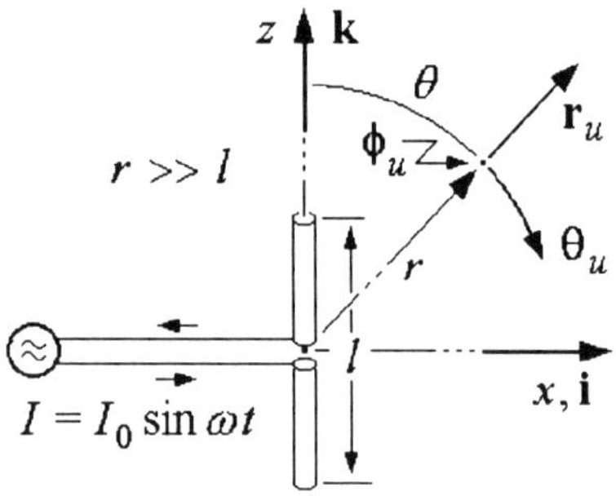

Fig. 2.2 Calculation of the electric and magnetic fields generated by an electric dipole antenna. (The unit vector ϕ_u is directed into the page.)

To find the magnetic field, we can use Eq. (2-2.5). Since the current in the antenna is filamentary, we can replace the volume integral in this equation by a line integral (note that for a filamentary current $\mathbf{J}dV' = \mathbf{J}sdl' = Id\mathbf{l}'$, where s is the cross-section area of the conductor, and $d\mathbf{l}'$ is a length element vector in the direction of $\mathbf{J}$). Furthermore, since the antenna is along the z axis, we can write Eq. (2-2.5) as

$$\mathbf{H} = \frac{1}{4\pi}\int\left\{\frac{[I]}{r^3} + \frac{1}{r^2 c}\left[\frac{\partial I}{\partial t}\right]\right\}\mathbf{k} \times \mathbf{r}\, dl' . \tag{2-2.21}$$

Differentiating Eq. (2-2.20), replacing t in Eq. (2-2.20) and in its derivative by the retarded time $t - r/c$, and substituting the resulting expressions in Eq. (2-2.21), we then have

$$\mathbf{H} = \frac{1}{4\pi}\int\left\{\frac{I_0 \sin\omega(t - r/c)}{r^3} + \frac{I_0\omega\cos\omega(t - r/c)}{r^2 c}\right\}\mathbf{k}\times\mathbf{r}\,dl'. \quad (2\text{-}2.22)$$

Since, by supposition, $r \gg \lambda = 2\pi c/\omega$, the first term in this integral is much smaller than the second term and can be neglected. Since $r \gg l$, r may be considered the same for all points of the antenna. The integral reduces therefore to the product of the second integrand and the length of the antenna

$$\mathbf{H} = \frac{I_0\omega\cos\omega(t - r/c)}{4\pi r^2 c}\mathbf{k}\times\mathbf{r}\,l, \quad (2\text{-}2.23)$$

or, in terms of the coordinates shown in Fig. 2.2,

$$\mathbf{H} = \frac{I_0 l\omega\cos\omega(t - r/c)}{4\pi r c}\sin\theta\,\boldsymbol{\phi}_u. \quad (2\text{-}2.24)$$

To find the electric field, we can use Eq. (2-2.12). Since we are only interested in the electric field at a large distance from the antenna, we can neglect in Eq. (2-2.12) all terms that approach zero at infinity faster than as $1/r$. We then have

$$\mathbf{E} = \frac{1}{4\pi\varepsilon_0 c^2}\int\left\{\frac{\mathbf{r}}{r^3}\left[\frac{\partial\mathbf{J}}{\partial t}\right]\cdot\mathbf{r} - \frac{1}{r}\left[\frac{\partial\mathbf{J}}{\partial t}\right]\right\}dV', \quad (2\text{-}2.25)$$

which we can write similar to Eq. (2-2.22) as

$$\mathbf{E} = \frac{1}{4\pi\varepsilon_0 c^2}\int\left\{\frac{\mathbf{r}(\mathbf{k}\cdot\mathbf{r})}{r^3}I_0\omega\cos\omega(t - r/c) - \frac{\mathbf{k}}{r}I_0\omega\cos\omega(t - r/c)\right\}dl'. \quad (2\text{-}2.26)$$

Taking into account that $\mathbf{k}\cdot\mathbf{r} = r\cos\theta$, and replacing the integral, as before, by the product of the integrand and the length of the antenna, we obtain

$$\mathbf{E} = \frac{lI_0\omega\cos\omega(t - r/c)}{4\pi\varepsilon_0 r c^2}\left(\frac{\mathbf{r}\cos\theta}{r} - \mathbf{k}\right). \quad (2\text{-}2.27)$$

Resolving $\mathbf{r}_u$ and $\boldsymbol{\theta}_u$ shown in Fig. 2.2 into components along the z and x axes, we can easily find that

$$\frac{\mathbf{r}\cos\theta}{r} - \mathbf{k} = \mathbf{r}_u\cos\theta - \mathbf{k} = \sin\theta\,\boldsymbol{\theta}_u . \tag{2-2.28}$$

Therefore we finally have

$$\mathbf{E} = \frac{lI_0\omega\cos\omega(t - r/c)}{4\pi\varepsilon_0 rc^2}\sin\theta\,\boldsymbol{\theta}_u . \tag{2-2.29}$$

An alternative method for obtaining Eq. (2-2.29) is to apply Maxwell's Eq. (2-1.4) to Eq. (2-2.24) and to integrate the result with respect to t.[8]

Example 2-2.3 Another system capable of generating electromagnetic waves is the radiating "magnetic dipole antenna," shown in Fig. 2.3. It consists of a circular loop of wire carrying a current

$$I = I_0\sin\omega t . \tag{2-2.30}$$

Assuming that the radius of the loop is a, find the electric and magnetic fields produced by this antenna at a large distance $r \gg \lambda = 2\pi c/\omega \gg a$ from it.

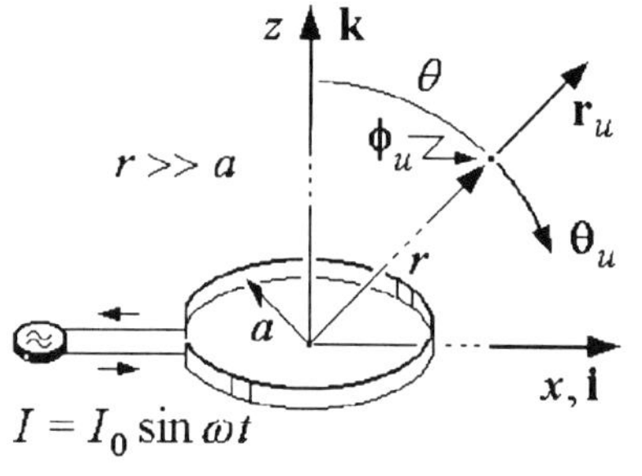

Fig. 2.3 Calculation of the electric and magnetic fields generated by a magnetic dipole antenna. (The unit vector $\boldsymbol{\phi}_u$ *is directed into the page.)*

We shall find the electric field produced by the antenna by using Eq. (2-2.4). Assuming that the antenna has no net charge, we only need to consider the second integral in this equation. Since the current in the antenna is filamentary, the volume integral can be replaced by a line integral (see Example 2-2.2). Differentiating then Eq. (2-2.30) and replacing t in the derivative by $t - r/c$, we can write Eq. (2-2.4) as

$$\mathbf{E} = -\frac{1}{4\pi\varepsilon_0 c^2}\oint \frac{I_0\omega\cos\omega(t - r/c)}{r}\,d\mathbf{l}', \qquad (2\text{-}2.31)$$

where $d\mathbf{l}'$ has the same direction as the current in the loop. Transforming the integral in Eq. (2-2.31) by means of vector identity (V-18), factoring out the constants, and using vector identity (V-25), we have

$$\begin{aligned}\mathbf{E} &= -\frac{I_0\omega}{4\pi\varepsilon_0 c^2}\int d\mathbf{S}' \times \nabla' \frac{\cos\omega(t - r/c)}{r} \\ &= +\frac{I_0\omega}{4\pi\varepsilon_0 c^2}\int\left\{-\frac{\omega}{rc}\sin\omega(t - r/c) + \frac{1}{r^2}\cos\omega(t - r/c)\right\}\mathbf{r}_u \times d\mathbf{S}'.\end{aligned} \qquad (2\text{-}2.32)$$

But $\omega/c = 2\pi/\lambda$ and, by the statement of the problem, $r \gg \lambda$. Therefore the second term in the last integral may be neglected, and we obtain

$$\mathbf{E} = -\frac{I_0\omega^2}{4\pi\varepsilon_0 c^3}\int \frac{\sin\omega(t - r/c)}{r}\mathbf{r}_u \times d\mathbf{S}'. \qquad (2\text{-}2.33)$$

Now, since $r \gg a$, we can replace the integral by the product of the integrand and the surface area of the antenna, so that

$$\mathbf{E} = -\frac{I_0\omega^2}{4\pi\varepsilon_0 c^3}\frac{\sin\omega(t - r/c)}{r}\mathbf{r}_u \times \mathbf{k}\,\pi a^2, \qquad (2\text{-}2.34)$$

or

$$\mathbf{E} = \frac{I_0\omega^2 a^2 \sin\omega(t - r/c)}{4\varepsilon_0 c^3 r}\sin\theta\,\boldsymbol{\phi}_u. \qquad (2\text{-}2.35)$$

The magnetic field can be determined from Eq. (2-2.5). Since we are only interested in the magnetic field at a large distance from the antenna, we can neglect in Eq. (2-2.5) the first term in the integrand (it is proportional to $1/r^2$, and for large r is negligible compared with the second term, which is proportional to $1/r$). We then have, replacing as before volume integration by line integration,

$$\mathbf{H} = -\frac{1}{4\pi}\oint \frac{I_0\omega\cos\omega(t - r/c)}{r^2c}\,\mathbf{r}\times d\mathbf{l}'. \tag{2-2.36}$$

Since $r \gg a$, r may be considered the same at all points of the antenna, and therefore we may factor out $\mathbf{r}/r$, obtaining

$$\mathbf{H} = -\frac{1}{4\pi cr}\,\mathbf{r}\times\oint \frac{I_0\omega\cos\omega(t - r/c)}{r}\,d\mathbf{l}'. \tag{2-2.37}$$

But the integral in Eq. (2-2.37) is the same as in Eq. (2-2.31) for **E**. By Eqs. (2-2.37) and (2-2.31)-(2-2.35), we then have

$$\mathbf{H} = \frac{I_0\omega^2a^2}{4c^2}\,\frac{\sin\omega(t - r/c)}{r^2}\sin\theta\,\mathbf{r}\times\boldsymbol{\phi}_u. \tag{2-2.38}$$

or

$$\mathbf{H} = -\frac{I_0\omega^2a^2}{4c^2}\,\frac{\sin\omega(t - r/c)}{r}\sin\theta\,\boldsymbol{\theta}_u. \tag{2-2.39}$$

▲

2-3. Surface Integrals for Retarded Electric and Magnetic Fields

A remarkable feature of Eqs. (2-2.1) and (2-2.2) is that they correlate the electric field with the *gradient* of the charge distribution and correlate the magnetic field with the *curl* of the current distribution rather than with the charge and current distribution as such. Hence, the equations may be interpreted as indicating that the electric and magnetic fields are associated not with electric charges and currents, but rather with the *inhomogeneities* in the distribution of charges and currents (a homogeneous, or uniform, charge distribution has zero gradient, and a homogeneous, or uniform, current distribution has zero curl).

A frequently encountered charge or current distribution is a distribution in which the charge or current changes abruptly from a finite value in the interior of the distribution to zero outside the distribution. For this type of charge and current distribution, Eqs. (2-2.1) and (2-2.2) can be transformed into special forms that are more convenient to use than Eqs. (2-2.1) and (2-2.2) themselves.

Consider first Eq. (2-2.1). In this equation the part of the integral involving $\nabla\rho$ can be separated into two integrals: the integral over the boundary layer of the charge distribution under consideration and the integral over the interior of the charge distribution:

$$\frac{1}{4\pi\varepsilon_0}\int\frac{[\nabla'\rho]}{r}dV' = \frac{1}{4\pi\varepsilon_0}\int_{B.l.}\frac{[\nabla'\rho]}{r}dV' + \frac{1}{4\pi\varepsilon_0}\int_{Int}\frac{[\nabla'\rho]}{r}dV'. \quad (2\text{-}3.1)$$

The first integral on the right of Eq. (2-3.1) can be transformed by using vector identity (V-33):

$$\frac{1}{4\pi\varepsilon_0}\int_{B.l.}\frac{[\nabla'\rho]}{r}dV' = \frac{1}{4\pi\varepsilon_0}\int_{B.l.}\nabla\frac{[\rho]}{r}dV' + \frac{1}{4\pi\varepsilon_0}\int_{B.l.}\nabla'\frac{[\rho]}{r}dV'. \quad (2\text{-}3.2)$$

In Eq. (2-3.2), ∇ in the first integral on the right operates upon the field point coordinates only. Therefore it can be factored out from under the integral sign. The integrand in this integral will then be $[\rho]/r$. Since both $[\rho]$ and r are finite, while the integration is over the volume of the boundary layer whose thickness, and therefore volume, can be assumed to be as small as we please, the integral vanishes. The second integral on the right of Eq. (2-3.2) can be transformed into a surface integral by using vector identity (V-20). Equation (2-3.2) can be written therefore as

$$\frac{1}{4\pi\varepsilon_0}\int_{B.\,layer}\frac{[\nabla'\rho]}{r}dV' = \frac{1}{4\pi\varepsilon_0}\oint_{B.layer}\frac{[\rho]}{r}d\mathbf{S}', \quad (2\text{-}3.3)$$

where the surface integral is extended over *both* surfaces (exterior and interior) of the boundary layer.

In Eq. (2-3.3), $d\mathbf{S}'$ of the exterior surface is directed into the space outside the charge distribution, while $d\mathbf{S}'$ of the interior surface is directed into the charge distribution. However, since there is no charge outside the charge distribution, the integral over the exterior surface vanishes. Since the boundary layer can be made as thin as we please, we can make the interior surface of the boundary layer coincide with the surface of the charge distribution. Reversing the sign in front of the surface integral, we can write then Eq. (2-3.3) as

$$\frac{1}{4\pi\varepsilon_0}\int_{B.\,layer}\frac{[\nabla'\rho]}{r}dV' = -\frac{1}{4\pi\varepsilon_0}\oint_{Boundary}\frac{[\rho]}{r}d\mathbf{S}', \qquad (2\text{-}3.4)$$

where the integration is now over the surface of the charge distribution, and where the surface element vector $d\mathbf{S}'$ is directed, as usual, from the charge distribution into the surrounding space.

From Eqs. (2-2.1), (2-3.1), and (2-3.4) we obtain

$$\mathbf{E} = \frac{1}{4\pi\varepsilon_0}\oint_{Boundary}\frac{[\rho]}{r}d\mathbf{S}' - \frac{1}{4\pi\varepsilon_0}\int_{Int}\frac{[\nabla'\rho]}{r}dV' - \frac{1}{4\pi\varepsilon_0 c^2}\int\frac{1}{r}\left[\frac{\partial\mathbf{J}}{\partial t}\right]dV'. \qquad (2\text{-}3.5)$$

This equation becomes especially simple in the case of a constant (uniform) charge distribution surrounded by a free space. In this case $\nabla\rho$ in the interior of the distribution is zero, and Eq. (2-3.5) simplifies to

$$\mathbf{E} = \frac{1}{4\pi\varepsilon_0}\oint_{Boundary}\frac{[\rho]}{r}d\mathbf{S}' - \frac{1}{4\pi\varepsilon_0 c^2}\int\frac{1}{r}\left[\frac{\partial\mathbf{J}}{\partial t}\right]dV'. \qquad (2\text{-}3.6)$$

Consider now Eq. (2-2.2). Just as in the case of Eq. (2-2.1), we can separate the integral in Eq. (2-2.2) into an integral over the boundary layer of the current distribution and an integral over the interior of the distribution. By the same reasoning as that used to simplify Eq. (2-3.2), we find that the integral over the boundary layer can be written as

$$\frac{1}{4\pi}\int_{B.layer}\frac{[\nabla'\times\mathbf{J}]}{r}dV' = \frac{1}{4\pi}\int_{B.layer}\nabla'\times\frac{[\mathbf{J}]}{r}dV'. \tag{2-3.7}$$

Transforming the integral on the right of Eq. (2-3.7) into a surface integral by means of vector identity (V-21), and taking into account that there is no current in the space outside the current distribution, we obtain, just as we obtained Eq. (2-3.4),

$$\frac{1}{4\pi}\int_{B.layer}\frac{[\nabla'\times\mathbf{J}]}{r}dV' = \frac{1}{4\pi}\oint_{Boundary}\frac{[\mathbf{J}]}{r}\times d\mathbf{S}', \tag{2-3.8}$$

where the integration is over the surface of the current distribution, and the surface element vector $d\mathbf{S}'$ is directed from the current distribution into the surrounding space.

Equation (2-2.2) can be written therefore as

$$\mathbf{H} = \frac{1}{4\pi}\oint_{Boundary}\frac{[\mathbf{J}]}{r}\times d\mathbf{S}' + \frac{1}{4\pi}\int_{Interior}\frac{[\nabla'\times\mathbf{J}]}{r}dV'. \tag{2-3.9}$$

For the special case of $\nabla\times\mathbf{J} = 0$ in the interior of the current distribution, Eq. (2-3.9) simplifies to

$$\mathbf{H} = \frac{1}{4\pi}\oint_{Boundary}\frac{[\mathbf{J}]}{r}\times d\mathbf{S}'. \tag{2-3.10}$$

▼

Example 2-3.1 A thin, uniformly charged disk of charge density ρ, radius a, and thickness b rotates with constant angular acceleration α about its axis, which is also the x axis of rectangular coordinates. The midplane of the disk coincides with the yz plane of the coordinates, and the rotation of the disk is right-handed relative to the x axis (Fig. 2.4). Using Eqs. (2-3.6) and (2-3.9), find the electric and magnetic fields produced by the disk at a point of the x axis, if at $t = 0$ the angular velocity of the disk is $\omega = 0$.

The disk creates a convection current $\mathbf{J} = \rho\mathbf{v} = \rho\omega R\boldsymbol{\theta}_u = \rho\alpha tR\boldsymbol{\theta}_u$, where R is the distance from the center of the disk, and $\boldsymbol{\theta}_u$ is a unit vector in the circular direction (right-handed with respect to α). The time derivative of $\mathbf{J}$ is $\partial\mathbf{J}/\partial t = \rho\alpha R\boldsymbol{\theta}_u$. To find $\nabla'\times\mathbf{J}$,

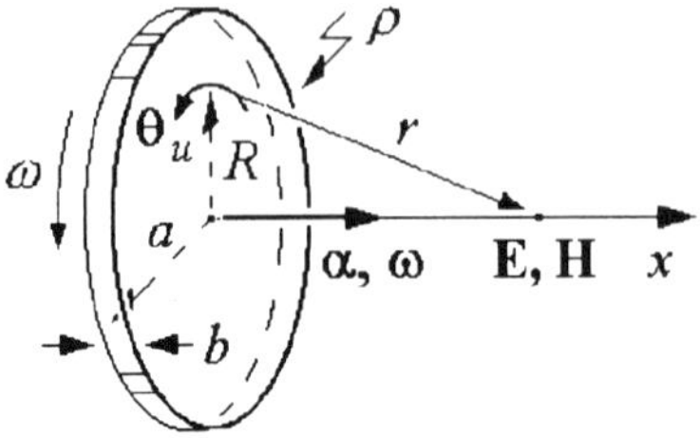

Fig. 2.4 Calculation of the electric and magnetic fields on the axis of a charged disk rotating with constant angular acceleration α.

we use the relation $\mathbf{v} = \boldsymbol{\omega} \times \mathbf{R}$ and vector identity (V-12). Taking into account that $\boldsymbol{\omega}$ is not a function of coordinates, we then obtain

$$\nabla' \times \mathbf{J} = \nabla' \times (\rho\boldsymbol{\omega} \times \mathbf{R}) = \rho[\boldsymbol{\omega}(\nabla' \cdot \mathbf{R}) - (\boldsymbol{\omega} \cdot \nabla')\mathbf{R}], \quad (2\text{-}3.11)$$

and since $\mathbf{R} = y'\mathbf{j} + z'\mathbf{k}$, while $\boldsymbol{\omega} \cdot \nabla' = \omega\partial/\partial x'$, we have

$$\nabla' \times \mathbf{J} = 2\rho\boldsymbol{\omega} = 2\rho\boldsymbol{\alpha}t = 2\rho\alpha t\mathbf{i}. \quad (2\text{-}3.12)$$

Examining now Eq. (2-3.6) and taking into account that $\partial\mathbf{J}/\partial t$ is in the circular direction, we recognize that the second integral in Eq. (2-3.6) vanishes by symmetry (see Example 2-2.1). And since ρ does not depend on time, we see from Eq. (2-3.6) that the electric field of the disk is the ordinary electrostatic field given by

$$\mathbf{E} = \frac{1}{4\pi\varepsilon_0}\oint_{Boundary}\frac{\rho}{r}d\mathbf{S}' = \frac{\rho}{4\pi\varepsilon_0}\oint_{Boundary}\frac{d\mathbf{S}'}{r}. \quad (2\text{-}3.13)$$

Let us now evaluate the surface integral in Eq. (2-3.13). By the symmetry of the system, only the two flat surfaces of the disk contribute to the field on the axis. The back surface is located at $x' = -b/2$, the front surface is located at $x' = +b/2$. The direction of the surface element vector $d\mathbf{S}'$ is $-\mathbf{i}$ for the back surface and $+\mathbf{i}$ for the front surface. We have therefore

$$\mathbf{E} = -\frac{\rho\mathbf{i}}{4\pi\varepsilon_0}\int_0^a\frac{2\pi R dR}{[R^2 + (x + b/2)^2]^{1/2}} + \frac{\rho\mathbf{i}}{4\pi\varepsilon_0}\int_0^a\frac{2\pi R dR}{[R^2 + (x - b/2)^2]^{1/2}}$$

$$= -\frac{\rho\mathbf{i}}{2\varepsilon_0}\{[a^2 + (x + b/2)^2]^{1/2} - (x + b/2) - [a^2 + (x - b/2)^2]^{1/2} + (x - b/2)\}. \quad (2\text{-}3.14)$$

Since $b \ll x$, we can use the relation

$$[a^2 + (x \pm b/2)^2]^{1/2} = [a^2 + x^2 \pm xb]^{1/2} = (a^2 + x^2)^{1/2}[1 \pm xb/2(a^2 + x^2)]. \tag{2-3.15}$$

Substituting Eq. (2-3.15) into Eq. (2-3.14), we obtain after elementary simplifications

$$\mathbf{E} = \frac{\rho b}{2\varepsilon_0}\left[1 - \frac{x}{(a^2 + x^2)^{1/2}}\right]\mathbf{i}. \tag{2-3.16}$$

To find the magnetic field, we use Eq. (2-3.9). Substituting [**J**] $= \rho\alpha R(t - r/c)\boldsymbol{\theta}_u$ and $[\boldsymbol{\nabla}' \times \mathbf{J}] = 2\rho\alpha(t - r/c)\mathbf{i}$ into Eq. (2-3.9), we have

$$\mathbf{H} = \frac{1}{4\pi}\oint_{Boundary}\frac{\rho\alpha R(t - r/c)}{r}\boldsymbol{\theta}_u \times d\mathbf{S}' + \frac{\mathbf{i}}{4\pi}\int_{Int}\frac{2\rho\alpha(t - r/c)}{r}dV'. \tag{2-3.17}$$

By the symmetry of the system, only the curved surface of the disk contributes to the first integral. At this surface $R = a$, $r = (a^2 + x^2)^{1/2}$, $\boldsymbol{\theta}_u \times d\mathbf{S}' = -\mathbf{i}\, dS'$, and the surface itself is $S' = 2\pi ab$. In the second integral r is $r = (R^2 + x^2)^{1/2}$ and the volume element is $dV' = b2\pi R dR$. The magnetic field is therefore

$$\mathbf{H} = -\mathbf{i}\frac{\rho\alpha a[t - (a^2 + x^2)^{1/2}/c^2]2\pi ab}{4\pi(a^2 + x^2)^{1/2}} + \frac{\mathbf{i}\rho\alpha}{2\pi}\int_0^a \frac{t - (R^2 + x^2)^{1/2}/c^2}{(R^2 + x^2)^{1/2}}2\pi bRdR$$

$$= -\mathbf{i}\frac{\rho\alpha t a^2 b}{2(a^2 + x^2)^{1/2}} + \mathbf{i}\frac{\rho\alpha a^2 b}{2c} + \mathbf{i}\rho\alpha t b(a^2 + x^2)^{1/2} - \mathbf{i}\frac{\rho\alpha a^2 b}{2c} \tag{2-3.18}$$

or

$$\mathbf{H} = \mathbf{i}\rho\alpha bt(a^2 + x^2)^{1/2}\left[1 - \frac{a^2}{2(a^2 + x^2)}\right]. \tag{2-3.19}$$

It is interesting to note that neither the electric nor the magnetic field of the rotating disk is retarded, just as was the case with the fields of the rotating ring discussed in Example 2-2.1 (see, however, Example 2-4.2).

▲

2-4. Retarded Potentials for Electric and Magnetic Fields

The calculation of time-dependent electric and magnetic fields can sometimes be simplified by using *retarded electromagnetic potentials*.

For the calculation of magnetic fields in a vacuum it is convenient to use the potentials defined in Corollary I of Section 1-1. Substituting in Eqs. (1-1.3), (1-1.4), and (1-1.5) $\mathbf{V} = \mathbf{B}$, $\nabla \cdot \mathbf{V} = \nabla \cdot \mathbf{B} = 0$, $K_1 = 0$, and $\mathbf{K}_2 = \mu_0\mathbf{J}$ [because by Eqs (2-1.12) and (1-1.1) $\mathbf{K} = \nabla \times \mathbf{J}$ in the wave equation for $\mathbf{H}$, so that $\mathbf{K} = \mu_0\nabla \times \mathbf{J}$ in the wave equations for $\mathbf{B} = \mu_0\mathbf{H}$], and leaving out, as usual, φ_0 and $\mathbf{A}_0$, we have

$$\mathbf{B} = \nabla \times \mathbf{A}, \tag{2-4.1}$$

where

$$\mathbf{A} = \frac{\mu_0}{4\pi}\int \frac{[\mathbf{J}]}{r} dV' . \tag{2-4.2}$$

If the current is filamentary, this equation reduces to

$$\mathbf{A} = \frac{\mu_0}{4\pi}\oint \frac{[I]}{r} d\mathbf{l}' , \tag{2-4.3}$$

where $d\mathbf{l}'$ is a length element vector in the direction of the current.

For the calculation of electric fields in a vacuum it is convenient to use the potentials defined in Corollary II of Section 1-1. Substituting in Eqs. (1-1.6), (1-1.7), and (1-1.8) $\mathbf{V} = \mathbf{E}$, $\nabla \cdot \mathbf{E} = \nabla \cdot (\mathbf{D}/\varepsilon_0) = \rho/\varepsilon_0$, $\mathbf{K} = -\mu_0\partial\mathbf{J}/\partial t$ [see Eq. (2-1.9)], and leaving out φ_0 and $\mathbf{W}_0$, we have

$$\mathbf{E} = -\nabla\varphi + \mathbf{W}, \tag{2-4.4}$$

where

$$\varphi = \frac{1}{4\pi\varepsilon_0}\int \frac{[\rho]}{r} dV' \tag{2-4.5}$$

while

$$\mathbf{W} = -\frac{\mu_0}{4\pi}\int \frac{1}{r}\left[\frac{\partial \mathbf{J}}{\partial t}\right] dV' . \tag{2-4.6}$$

Using Eq. (1-2.14) and taking into account that the integration in Eq. (2-4.6) involves space coordinates only, we can factor out $\partial/\partial t$ from under the integral sign, obtaining

$$-\frac{\mu_0}{4\pi}\int\frac{1}{r}\left[\frac{\partial \mathbf{J}}{\partial t}\right]dV' = -\frac{\partial}{\partial t}\left\{\frac{\mu_0}{4\pi}\int\frac{[\mathbf{J}]}{r}dV'\right\}. \tag{2-4.7}$$

Therefore, according to Eq. (2-4.2), Eq. (2-4.4) can be written as

$$\mathbf{E} = -\nabla\varphi - \frac{\partial \mathbf{A}}{\partial t}, \tag{2-4.8}$$

where **A** is the retarded magnetic vector potential given by Eq. (2-4.2) or Eq. (2-4.3).

The potentials φ and **A** given by Eqs. (2-4.5) and (2-4.2) are the retarded electromagnetic potentials. They represent a generalization of the ordinary electric and magnetic potentials φ and **A** and reduce to them in the case of time-independent fields in a vacuum.[9]

▼

Example 2-4.1 Show that the retarded potentials φ and **A** satisfy *Lorentz's condition*

$$\nabla\cdot\mathbf{A} = -\varepsilon_0\mu_0\frac{\partial\varphi}{\partial t}. \tag{2-4.9}$$

From Eqs. (2-4.5) and (1-2.14) we have

$$-\varepsilon_0\mu_0\frac{\partial\varphi}{\partial t} = -\frac{\mu_0}{4\pi}\int\frac{\partial}{\partial t}\frac{[\rho]}{r}dV' = -\frac{\mu_0}{4\pi}\int\frac{1}{r}\left[\frac{\partial\rho}{\partial t}\right]dV'. \tag{2-4.10}$$

But according to the continuity law, Eq. (2-2.6),

$$-\left[\frac{\partial\rho}{\partial t}\right] = [\nabla'\cdot\mathbf{J}], \tag{2-4.11}$$

so that

$$-\varepsilon_0\mu_0\frac{\partial\varphi}{\partial t} = \frac{\mu_0}{4\pi}\int\frac{[\nabla'\cdot\mathbf{J}]}{r}dV'. \tag{2-4.12}$$

Transforming the integral in Eq. (2-4.12) by means of vector identity (V-27), we have

$$-\varepsilon_0\mu_0\frac{\partial\varphi}{\partial t} = \frac{\mu_0}{4\pi}\int\nabla\cdot\frac{[\mathbf{J}]}{r}dV' + \frac{\mu_0}{4\pi}\int\nabla'\cdot\frac{[\mathbf{J}]}{r}dV'. \qquad (2\text{-}4.13)$$

The last integral can be transformed into a surface integral by means of the vector identity (V-19), and since there is no current at infinity, the surface integral is zero, and so is the last integral. In the first integral, ∇ can be factored out from under the integral sign. Therefore we obtain

$$-\varepsilon_0\mu_0\frac{\partial\varphi}{\partial t} = \nabla\cdot\frac{\mu_0}{4\pi}\int\frac{[\mathbf{J}]}{r}dV'. \qquad (2\text{-}4.14)$$

Eliminating the last integral in Eq. (2-4.14) by means of Eq. (2-4.2), we obtain Lorentz's condition.

Example 2-4.2 Find the electric and magnetic fields at all points of space far from the rotating ring described in Example 2-2.1 (Fig. 2.5).

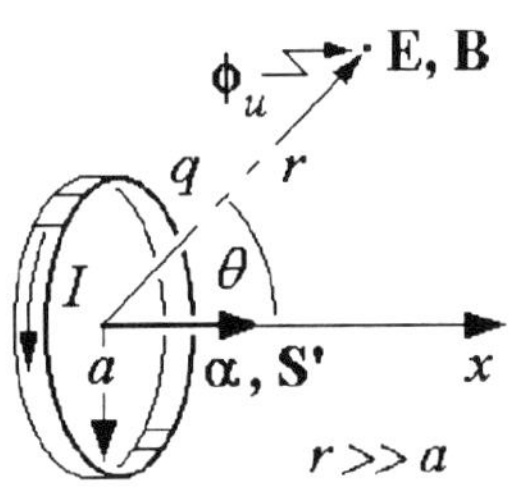

Fig. 2.5 Calculation of the electric and magnetic fields far from the charged ring rotating with constant angular acceleration. (The unit vector ϕ_u is directed into the page.)

At large distances from the ring, the ring constitutes a point charge q, which does not depend on time. Therefore the electric potential of the ring is the ordinary electrostatic potential

$$\varphi = \frac{q}{4\pi\varepsilon_0 r}. \qquad (2\text{-}4.15)$$

Since the ring constitutes a convection line current $I = q\alpha t/2\pi$, the magnetic vector potential of the ring is, by Eq. (2-4.3),

$$\mathbf{A} = \frac{\mu_0}{4\pi}\oint \frac{q\alpha(t-r/c)/2\pi}{r}\,d\mathbf{l}' = \frac{q\alpha t\mu_0}{8\pi^2}\oint \frac{d\mathbf{l}'}{r} - \frac{q\alpha\mu_0}{8\pi^2 c}\oint d\mathbf{l}' . \qquad (2\text{-}4.16)$$

The last integral on the right of Eq. (2-4.16) is zero. The remaining integral can be transformed into a surface integral by means of vector identity (V-18). We then obtain

$$\mathbf{A} = \frac{q\alpha t\mu_0}{8\pi^2}\oint \frac{d\mathbf{l}'}{r} = \frac{q\alpha t\mu_0}{8\pi^2}\int \frac{\mathbf{r}'_u}{r^2}\times d\mathbf{S}', \qquad (2\text{-}4.17)$$

where $\mathbf{r}'_u$ is a unit vector directed from the point of observation toward the surface element $d\mathbf{S}'$.

Now, since the point of observation is far from the ring, the integral can be replaced by the (vector) product of the integrand and the surface area $\mathbf{S}'$ of the ring, so that the vector potential is

$$\mathbf{A} = \frac{q\alpha t\mu_0}{8\pi^2 r^2}\mathbf{r}'_u\times\mathbf{S}' = -\frac{q\alpha t\mu_0}{8\pi^2 r^2}\mathbf{r}_u\times\mathbf{S}', \qquad (2\text{-}4.18)$$

where $\mathbf{r}_u$ is a unit vector directed from the ring toward the point of observation.The magnitude of the vector $\mathbf{S}'$ is πa^2, and the direction is along the x axis. Designating the angle between $\mathbf{r}_u$ and $\mathbf{S}'$ as θ, we then have for the vector potential

$$\mathbf{A} = -\frac{qa^2\alpha t\mu_0}{8\pi r^2}\sin\theta\,\boldsymbol{\varphi}_u, \qquad (2\text{-}4.19)$$

where $\boldsymbol{\varphi}_u$ is a unit vector in the circular direction left-handed relative to the x axis.

By Eq. (2-4.1), the magnetic flux density field associated with this vector potential is

$$\mathbf{B} = \nabla\times\mathbf{A} = \frac{qa^2\alpha t\mu_0}{8\pi r^3}(2\cos\theta\,\mathbf{r}_u + \sin\theta\,\boldsymbol{\theta}_u) \qquad (2\text{-}4.20)$$

(we do not reproduce the actual calculation of $\nabla \times \mathbf{A}$, since it is not important for the purpose of the present example; the calculation is done by using the expressions for the curl of a vector in spherical coordinates[10]). It is interesting to note that this field is an ordinary (unretarded) field of a current dipole,[11] and that on the x axis ($\theta = 0$) it reduces to the field found in Example 2-2.1 (for $x \gg a$).

Let us now find the electric field of the ring. By Eq. (2-4.8), (2-4.15), and (2-4.19), we have

$$\mathbf{E} = \frac{q}{4\pi\varepsilon_0 r^2}\mathbf{r}_u + \frac{qa^2\alpha\mu_0}{8\pi r^2}\sin\theta\,\boldsymbol{\varphi}_u \tag{2-4.21}$$

or, using $\varepsilon_0\mu_0 = 1/c^2$,

$$\mathbf{E} = \frac{q}{4\pi\varepsilon_0 r^2}\mathbf{r}_u + \frac{qa^2\alpha}{8\pi\varepsilon_0 c^2 r^2}\sin\theta\,\boldsymbol{\varphi}_u. \tag{2-4.22}$$

It is interesting to note that although the electric field of the ring does not depend on t, the presence of the φ_u term makes the field different from the electrostatic field of the ring. This term represents the contribution of $[\partial\mathbf{J}/\partial t]$ in Eq. (2-2.1) and represents the "electrokinetic field" (see Section 2-5). In the case under consideration, the electrokinetic field is circular and is directed opposite to the current in the ring.

On the x axis, the electric field of the ring reduces to the field found in Example 2-2.1.

▲

2-5. Electromagnetic Induction

Electromagnetic induction is frequently explained as a phenomenon in which a changing magnetic field produces an electric field ("Faraday induction") and a changing electric field produces a magnetic field ("Maxwell induction").

A detailed examination of the causal relations in time-dependent electric and magnetic fields shows, however, that neither of the two fields can create the other.[12] The causal equations for electric and magnetic fields in a vacuum are the retarded field equations discussed in Sections 2-2 and 2-3.

According to Eqs. (2-2.1), (2-2.2), (2-2.4), (2-2.5), and (2-2.12), in time-variable systems electric and magnetic fields are always created simultaneously, because they have a common causative source: the changing electric current $\partial\mathbf{J}/\partial t$. Once created, the two fields coexist from then on without any effect upon each other. Therefore electromagnetic induction as a phenomenon in which one of the fields creates the other is an illusion. The illusion of the "mutual creation" arises from the fact that in time-dependent systems the two fields always appear prominently together, while their causative sources (the time-variable current in particular) remain in the background.

As can be seen from Eq. (2-2.1) or from Eq. (2-2.4), a time-variable electric current creates an electric field parallel to that current (parallel to $[\partial\mathbf{J}/\partial t]$). This field exerts an electric force on the charges in nearby conductors thereby creating induced electric currents in the conductors. Thus, the term "electromagnetic induction" is actually a misnomer, since no magnetic effect is involved in the phenomenon, and since the induced current is caused solely by the time-variable electric current and by the electric field produced by that current.

The electric field produced by a time-variable current differs in two important respects from the ordinary electric field produced by electric charges at rest: first, the field produced by a current is directed along the current rather than along a radius vector, and second, the field exists only as long as the current is changing in time. Therefore the electric force caused by this field is also different from the ordinary electric (electrostatic) force: it is directed along the current and it lasts only as long as the current is changing. Unlike the electrostatic force, which is always an attraction or repulsion between electric charges, the electric force due to a time-variable current is a *dragging* force: it causes electric charges to move parallel (or anti-parallel) relative to the direction of the current. If the time-variable current is a convection current, then the force that this current exerts on neighboring charges causes them to move parallel to the convection current, rather than toward or away from the charges forming the convection current [the total force is, of course, given by all the terms in Eq. (2-2.1) or Eq. (2-2.4)].

Since the electric field created by time-variable currents is very different from all other fields encountered in electromagnetic phenomena, a special name should be given to it. Taking into account that the cause of this field is a motion of electric charges (current), we may call it the *electrokinetic field*, and we may call the force which this field exerts on an electric charge the *electrokinetic force*.[13] Of course, we could simply call this field the "induced field." However, such a name would not reflect the special nature and properties of this field.

Let us designate the electrokinetic field by the vector $\mathbf{E}_k$. From Eq. (2-2.4) we thus have

$$\mathbf{E}_k = -\frac{1}{4\pi\varepsilon_0 c^2}\int \frac{1}{r}\left[\frac{\partial \mathbf{J}}{\partial t}\right] dV'. \tag{2-5.1}$$

The electrokinetic field provides a precise and clear explanation of one of the most remarkable properties of electromagnetic induction: Lenz's law. Consider a straight current-carrying conductor parallel to another conductor. According to Lenz's law, the current induced in the second conductor is opposite to the inducing current in the first conductor when the inducing current is increasing, and is in the same direction as the inducing current when the inducing current is decreasing. In the past no convincing explanation of this effect was known. But the electrokinetic field provides the definitive explanation of Lenz's law: by Eq. (2-5.1), the sign (direction) of the electrokinetic field is opposite to the sign of the time derivative of the *inducing* current. When the derivative is positive, the electrokinetic field is opposite to the inducing current; when the derivative is negative, the electrokinetic field is in the same direction as the inducing current. Since the *induced* current is caused by the electrokinetic field, the direction of this field determines the direction of the induced current: opposite to the inducing current when that current increases (positive derivative), the same as the inducing current when the inducing current decreases (negative derivative).

Of course, since the direction of the inducing current usually varies from point to point in space, the ultimate direction of the electrokinetic field and of the current that it produces is determined, in general, by the combined effect of all the current elements of the inducing current in the integral of Eq. (2-5.1).

The electrokinetic field also gives a simple explanation of the fact (first noted by Faraday) that the strongest induced current is produced between parallel conductors, whereas no induction takes place between conductors at right angles to each other. This phenomenon is now easily understood from the fact that the electrokinetic field due to a straight conductor carrying an inducing current is always parallel to the conductor.

Although we have been discussing the electrokinetic field as the cause of induced currents in conductors, its significance is much more general. This field can exist anywhere in space and can manifest itself as a pure force field by its action on free electric charges. Of course, because of the c^2 in the denominator in Eq. (2-5.1), the electrokinetic field cannot be particularly strong except when the current changes very fast. This is probably the main reason why this field was ignored in the past. Another reason is the temporal (transient) nature of this field.

But even weak electric fields can produce strong currents in conductors, and that is why the current-producing effect of the electrokinetic field is much more prominent than its force effect on electric charges in free space.

If we compare Eq. (2-5.1) with Eq. (2-4.2) for the retarded magnetic vector potential **A** produced by a current **J**, we recognize that the electrokinetic field is equal to the negative time derivative of **A** (observe that $\mu_0 = 1/\varepsilon_0 c^2$):

$$\mathbf{E}_k = -\frac{\partial \mathbf{A}}{\partial t}. \tag{2-5.2}$$

However, although Eq. (2-5.2) correlates the electrokinetic field with the magnetic vector potential, there is no causal link between the two: the correlation merely reflects the fact that both the electrokinetic field and the magnetic vector potential are simultaneously caused by the same electric current.

Important as it is, the electrokinetic field has not been studied (or even recognized as a special force field) until very recently, although the fact that the time derivative of the retarded vector potential is associated with an electric field has been known for a long time.

Electromagnetic induction is a phenomenon associated with relatively slow current variations and with electromagnetic fields extending over relatively small regions of space (rapid current variations and time-variable fields extending over long distances are dealt with on the basis of radiation theory; see Examples 2-2.2 and 2-2.3). More specifically, electromagnetic induction applies to systems satisfying Eq. (2-2.13). Therefore, as far as electromagnetic induction is concerned, the retardation in the propagation of the electric field from the inducing current to the conductor in which the induced current is created can be ignored. Removing the retardation symbol [ ] in Eq. (2-5.1) and factoring out $\partial/\partial t$, we then obtain for the electrokinetic field

$$\mathbf{E}_k = -\frac{\partial}{\partial t}\left(\frac{1}{4\pi\varepsilon_0 c^2}\int \frac{\mathbf{J}}{r}\,dV'\right). \tag{2-5.3}$$

▼

Example 2-5.1 A conducting circular ring of radius R is placed outside a long coaxial solenoid of n turns, radius a and length L, carrying a current I (Fig. 2.6). Using Eq. (2-5.3) find the electrokinetic field and then the voltage induced in the ring when the current in the solenoid is changing. Observe that according to the conventional explanation of electromagnetic induction, the voltage and the current in the ring is induced by the changing magnetic field at the location of the ring. But this explanation does not work in the present case, because there is no magnetic field at the location of the ring (except for the end-effect field of the solenoid, which is negligible).

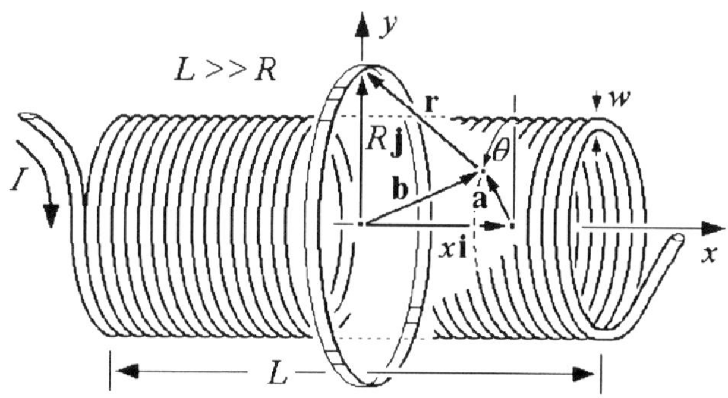

Fig. 2.6 Calculation of the voltage induced in a conducting ring placed outside a solenoid carrying a variable current.

Let the axis of the solenoid be the x axis of a rectangular system of coordinates, let the ring be in the yz plane, and let the ends of the solenoid be at $x = -L/2$ and $x = L/2$. To find the electrokinetic field induced by the solenoid in the ring, we shall consider a point of the ring located on the y axis. We can represent this point by the vector $R\mathbf{j}$. Consider next a point on the surface of the solenoid at a distance x from the yz plane. Combining cylindrical and rectangular coordinates, we can represent that point by the vector $\mathbf{b} = x\mathbf{i} + a\cos\theta\mathbf{j} + a\sin\theta\mathbf{k}$. The distance between the two points is then $\mathbf{r} = R\mathbf{j} - \boldsymbol{b} = -x\mathbf{i} + (R - a\cos\theta)\,\mathbf{j} - a\sin\theta\mathbf{k}$, so that for r in Eq. (2-5.3) we have, by adding the squares of $\mathbf{r}$ and taking the square root of the sum, $r = (R^2 + a^2 - 2Ra\cos\theta + z^2)^{1/2}$. The current density in the solenoid can be written as

$\mathbf{J} = nI/Lw\boldsymbol{\theta}_u = nI/Lw(-\sin\theta\mathbf{j} + \cos\theta\mathbf{k})$, where I is the current in the solenoid, w is the thickness of the current sheet, and $\boldsymbol{\theta}_u$ is a unit vector in the direction of the current. The volume element to be used in Eq. (2-5.3) can be written as $dV' = wad\theta dx$.

By the symmetry of the system, the contribution of the y component of $\mathbf{J}$ to the electrokinetic field is zero. Equation (2-5.3) becomes therefore (we replace $1/\varepsilon_0 c^2$ by μ_0)

$$\mathbf{E}_k = -\frac{\partial}{\partial t}\left(\frac{\mu_0}{4\pi}\int_0^{2\pi}\int_{-L/2}^{L/2}\frac{nI\cos\theta\mathbf{k}}{Lw(R^2+a^2-2Ra\cos\theta+x^2)^{1/2}}wad\theta dx\right), \quad (2\text{-}5.4)$$

or

$$\mathbf{E}_k = -\frac{\partial}{\partial t}\left(\mathbf{k}\frac{\mu_0 nIa}{4\pi L}\int_0^{2\pi}\int_{-L/2}^{L/2}\frac{\cos\theta}{(R^2+a^2-2Ra\cos\theta+x^2)^{1/2}}d\theta dx\right). \quad (2\text{-}5.5)$$

Integrating by parts over θ, we obtain

$$\mathbf{E}_k = -\frac{\partial}{\partial t}\left(\mathbf{k}\frac{\mu_0 nIRa^2}{4\pi L}\int_0^{2\pi}\int_{-L/2}^{L/2}\frac{\sin^2\theta}{(R^2+a^2-2Ra\cos\theta+x^2)^{3/2}}d\theta dx\right). \quad (2\text{-}5.6)$$

Integrating over x and taking into account that $L \gg R$, a, we obtain

$$\mathbf{E}_k = -\frac{\partial}{\partial t}\left(\mathbf{k}\frac{\mu_0 nIRa^2}{2\pi L}\int_0^{2\pi}\frac{\sin^2\theta}{(R^2+a^2-2Ra\cos\theta)}d\theta\right). \quad (2\text{-}5.7)$$

The integral in Eq. (2-5.7) is just[14] π/R^2. The electrokinetic field generated at the point $R\mathbf{j}$ of the ring by the current in the solenoid is therefore (replacing $\mathbf{k}$ by $\boldsymbol{\theta}_u$)

$$\mathbf{E}_k = -\frac{\partial}{\partial t}\left(\boldsymbol{\theta}_u\frac{\mu_0 nIa^2}{2RL}\right), \quad (2\text{-}5.8)$$

and the voltage induced in the ring is

$$V_{ind} = \oint \mathbf{E_k}\cdot d\mathbf{l} = E_k 2\pi R = -\mu_0\frac{n\pi a^2}{L}\frac{\partial I}{\partial t}. \quad (2\text{-}5.9)$$

▲

References and Remarks for Chapter 2

1. See, for example, Oleg D. Jefimenko, *Electricity and Magnetism*, 2nd ed., (Electret Scientific, Star City, 1989) pp. 500-503.

2. These equations constitute time-dependent counterparts of the corresponding time-independent equations. See Ref. 1, pp. 101-103 and 350-352.

3. For alternative derivations of Eqs. (2-2.4) and (2-2.5) see David J. Griffiths and Mark A. Heald, "Time-dependent generalization of the Biot-Savart and Coulomb laws," Am. J. Phys. **59**, 11-117(1991) and Tran-Cong Ton, "On the time-dependent, generalized Coulomb and Biot-Savart laws," Am. J. Phys. **59**, 520-528 (1991).

4. See, for example, Ref. 1, p. 497.

5. A similar equation is derived in W. K. H. Panofsky and M. Phillips, *Classical Electricity and Magnetism*, 2nd ed., (Addison-Wesley, Reading, 1962) p. 248.

6. See, for example, Ref. 1, pp. 99-100.

7. See, for example, Ref. 1, pp. 346-347.

8. For a more detailed solution see Ref. 1, pp. 559-562.

9. Equations (2-4.5) and (2-4.2) were first obtained in 1867 by L. Lorenz. See E. T. Whittaker, *A History of the Theories of Aether and Electricity* (Thomas Nelson, London, 1953) Vol. I, Chapt. 8 ("Maxwell") pp. 267-268.

10. See, for example, Ref. 1, p. 55.

11. See, for example, Ref. 1, pp. 380-382.

12. See Oleg D. Jefimenko, *Causality, Electromagnetic Induction, and Gravitation* (Electret Scientific, Star City, 1992) pp. 3-18.

13. The term "electrokinetic" is also used in reference to phenomena associated with the movement of charged particles through a continuous medium or with the movement of a continuous medium over a charged surface. These phenomena have no connection with the electrokinetic field discussed in this book.

14. It is best evaluated by using a computer program for symbolic integration.

3

RETARDED INTEGRALS FOR ELECTRIC AND MAGNETIC FIELDS AND POTENTIALS OF MOVING CHARGES

In this chapter we shall learn how retarded integrals for electric and magnetic fields and potentials can be used for finding electric and magnetic fields and potentials of moving electric charge distributions. We shall also discover important relations between the electric and magnetic fields for two special cases of moving charge distributions: an arbitrary charge distribution moving with constant velocity and a point charge in arbitrary motion.

3-1. Using Retarded Integrals for Finding Electric and Magnetic Fields and Potentials of Moving Charge Distributions

A time-variable electric charge distribution always involves a movement of electric charges. For example, if the density of a charge distribution changes with time, then some electric charges change their location within the charge distribution or move to or from the charge distribution. Conversely, a moving charge distribution is inevitably a time-variable charge distribution because it creates charge density in regions of space which it enters and eliminates charge density from the regions of space which it leaves. Consequently, the electric and magnetic fields of a moving charge distribution can be determined from retarded field (or retarded potential) equations derived in Chapter 2 for the general case of time-dependent charge and current distributions.

To use retarded field integrals for finding electric and magnetic fields of moving charge distributions, we need to express the time derivatives $\partial\rho/\partial t$ and $\partial\mathbf{J}/\partial t$ in terms of the velocity of the charge distribution under consideration. This can be done as follows. Consider a stationary charge distribution of density ρ as a function of x', y', z',

$$\rho = \rho(x',y',z'). \tag{3-1.1}$$

When this charge distribution moves with velocity $\mathbf{v}$ without changing its composition, the total time derivative of ρ is

$$\frac{d\rho}{dt} = \frac{\partial\rho}{\partial t} + \frac{\partial\rho}{\partial x'}\frac{dx'}{dt} + \frac{\partial\rho}{\partial y'}\frac{dy'}{dt} + \frac{\partial\rho}{\partial z'}\frac{dz'}{dt} = \frac{\partial\rho}{\partial t} + \mathbf{v}\cdot\nabla'\rho. \tag{3-1.2}$$

Since ρ remains the same as the charge moves, $d\rho/dt = 0$, so that

$$\frac{\partial\rho}{\partial t} = -\mathbf{v}\cdot\nabla'\rho. \tag{3-1.3}$$

A moving charge distribution constitutes a current whose density is $\mathbf{J} = \rho\mathbf{v}$. Therefore

$$\frac{\partial\mathbf{J}}{\partial t} = \frac{\partial(\rho\mathbf{v})}{\partial t} = -(\mathbf{v}\cdot\nabla'\rho)\mathbf{v} + \rho\frac{\partial\mathbf{v}}{\partial t} = -(\mathbf{v}\cdot\nabla'\rho)\mathbf{v} + \rho\dot{\mathbf{v}}. \tag{3-1.4}$$

Observe that in the retarded field integrals derived in Chapter 2, the denominator r representing the distance between the volume element dV' and the point of observation is not a function of time. Therefore it is not a function of time also in the case of moving charge distributions. A moving charge distribution must be considered as moving past different volume elements of space associated with different but fixed r's. The question arises, if dV'

is a volume element of space, rather than a volume element of a moving charge distribution, how does one introduce the volume of the charge distribution into the field integrals? To answer this question, let us examine how the electric and magnetic fields of a moving charge distribution are created.

The phenomenon of retardation indicates that time-dependent charge distributions send out electric (and magnetic) field "signals" that propagate in all directions with the velocity of light. The electric or magnetic field created by a time-variable charge distribution at the point of observation is the result of the signals sent out by all the individual charges within the distribution and simultaneously "received" at the point of observation at the instant t. But different charges within the distribution are at different distances from the point of observation, and the times needed for the signals originating from the different charges to arrive at the point of observation are different. Therefore the signals that are received at the point of observation simultaneously at the instant t are sent out from the different charges within the distribution at different retarded times $t' = t - r/c$. For a moving charge distribution these times are different not only because different charges within the distribution are located at different distances from the point of observation, but also because the location of these charges changes as the charge distribution moves. As a result, the region of space from which the field signals responsible for the field at the point of observation are sent is not equal to the region of space, or volume, occupied by the charge distribution when it is at rest.

Consider a charge distribution of length l moving against the x axis with a constant velocity $\mathbf{v}$. The electric field $\mathbf{E}$ of the charge is observed at the point O (Fig. 3.1). A field signal is sent from the trailing end of the distribution when this end is at the distance r_1 from the point of observation. A field signal is sent from the leading end, when this end is at the distance r_2 from the point of observation. Since the leading end is closer to the point of observation than the trailing end, the field signal from the leading end must be sent at a later time, if it is to arrive at the point of observation simultaneously with the signal sent from the trailing end. The difference in the times needed for the two signals to arrive at the point of observation is $r_1/c - r_2/c$. During this time the charge distribution moves a distance $(r_1/c - r_2/c)v$. Hence the distance l^* between the two points from which the two signals are sent is

$$l^* = (r_1 - r_2)v/c + l. \qquad (3\text{-}1.5)$$

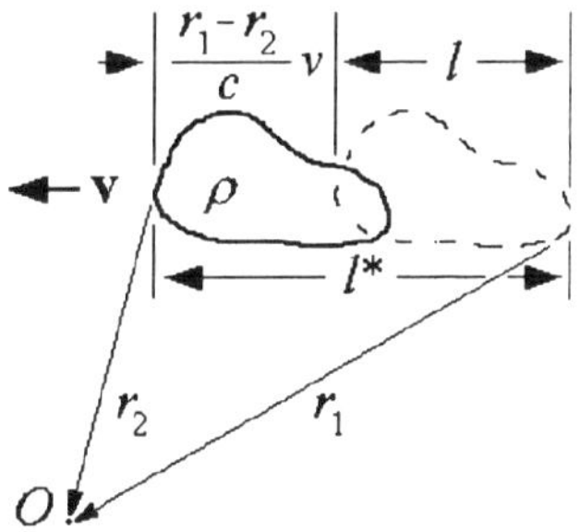

Fig. 3.1 *For the two field signals to arrive simultaneously at O, the field signal originating from the leading end of the moving charge must be sent later than the field signal originating from the trailing end of the charge.*

In this chapter we shall be mainly concerned with the special case of charge distributions for which $r_1, r_2 \gg l^*$. In this case, $r_1 - r_2 = l^* \cos\phi = l^*(\mathbf{r}\cdot\mathbf{v})/rv$, where r is the distance between the midpoint of l^* and the point of observation, and ϕ is the angle between $\mathbf{r}$ and $\mathbf{v}$ as shown in Fig. 3.2. Substituting this expression for $r_1 - r_2$ in Eq. (3-1.5), we have

$$l^* = l^*(\mathbf{r}\cdot\mathbf{v})/rc + l, \qquad (3\text{-}1.6)$$

or

$$l^* = \frac{l}{1-(\mathbf{r}\cdot\mathbf{v})/rc}. \qquad (3\text{-}1.7)$$

Therefore, as already mentioned, the region of space from which

Fig. 3.2 *Geometrical relations between r, θ, and l* when r_1, $r_2 \gg l^*$. The significance of the vector* **l*** *will be explained later.*

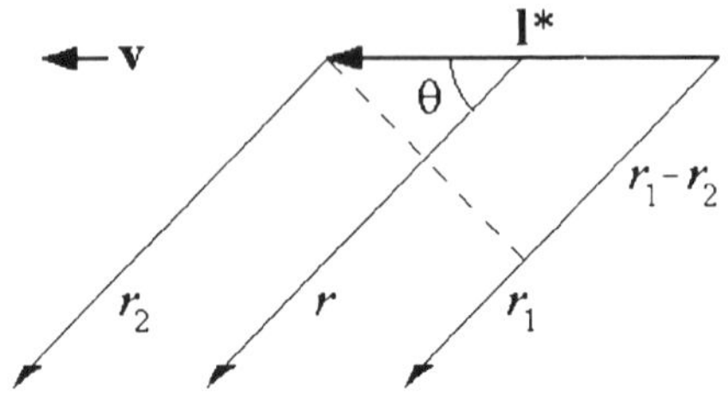

the moving charge sends out the field signals resulting in the electric and magnetic fields created at the point of observation is not equal to the region of space (volume) actually occupied by the charge. In the case of a charge distribution whose linear dimensions are small compared with the distance from the charge to the point of observation, this region of space, usually called the *effective volume*, or the *retarded volume*, $\Delta V'_{ret}$ is

$$\Delta V'_{ret} = \frac{\Delta V'}{1 - (\mathbf{r} \cdot \mathbf{v})/rc}, \tag{3-1.8}$$

where $\Delta V'$ is the actual volume of the charge [this equation is obtained from Eq. (3-1.7) by noting that the volume dimensions perpendicular to the direction of motion are not affected by retardation, and that the dimensions along the direction of motion change in accordance with Eq. (3-1.7)].

Although the distance l^* given by Eq. (3-1.5) or Eq. (3-1.7) is a distance between two points in space rather than a length of an object, it is usually called the *retarded length* of the charge. In fact, it is actually the "visual" length of a rapidly moving body, as the length of the body would appear to a stationary observer. As follows from Eq. (3-1.7), the retarded length of a body moving toward the observer is longer, and the retarded length of a body moving away from the observer is shorter, than the actual length of the body.[1] It should be emphasized that Eqs. (3-1.6)-(3-1.8) hold only for charges or bodies observed from a distance much greater than the linear dimensions of the charge or body. For a general case, the retarded length or volume of a body cannot be expressed by a simple formula, but can be calculated in terms of the actual length of the body once the position of the body at the time of observation is given (Section 4-3).

Another effect of retardation that needs to be taken into account when applying retarded field equations to moving charge distributions is an apparent distortion of the shape of a moving charge distribution. The distribution appears to change its shape because the retarded times for different points within the distribution are different.

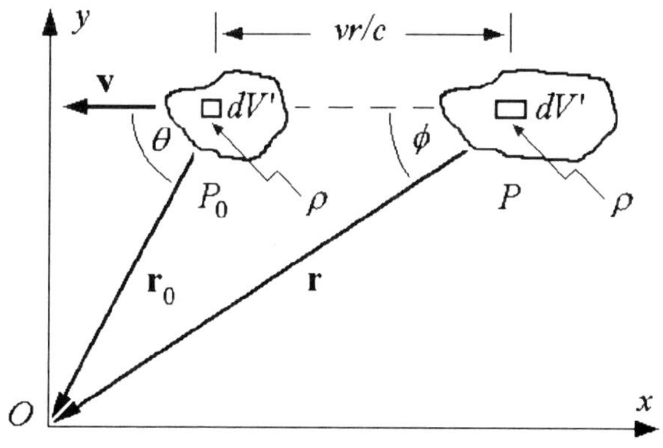

Fig. 3.3 Geometrical relations between the "present position vector" $\mathbf{r}_0$ *and the "retarded position vector"* $\mathbf{r}$ *for a charge distribution moving with velocity* $\mathbf{v}$ *in the negative x direction.*

Consider a charge distribution moving against the x axis with a velocity $\mathbf{v}$ and observed from a point O (Fig. 3.3). The retarded volume element dV' of the charge distribution is at the point P and is represented by the vector $\mathbf{r}$. The present position of the same volume element is at the point P_0 and is represented by the vector $\mathbf{r}_0$. The distance $\Delta x'$ from P to P_0 is the distance that the charge travels during the time that it takes the field signal to propagate from P to O, that is, $\Delta x' = v(r/c)$. We shall now show that, within the charge, any line parallel to the y axis when the charge is at rest or at its present position appears to be slanted when the charge is moving and is at a retarded position.

First, let us note that according to Fig. 3.3 the relation between the x component of the present position vector $\mathbf{r}_0$ and the x and y components of the retarded position vector $\mathbf{r}$ is (as usual, we use primes to indicate source-point coordinates)

$$x' = x'_0 + vr/c, \tag{3-1.9}$$

or

$$x' = x'_0 + (x'^2 + y'^2)^{1/2} v/c. \tag{3-1.10}$$

Differentiating Eq. (3-1.10) while keeping x'_0 constant, we have

$$\frac{dx'}{dy'} = \frac{y'(v/c)}{r[1 - (v/c)(x'/r)]}, \tag{3-1.11}$$

which can be written as

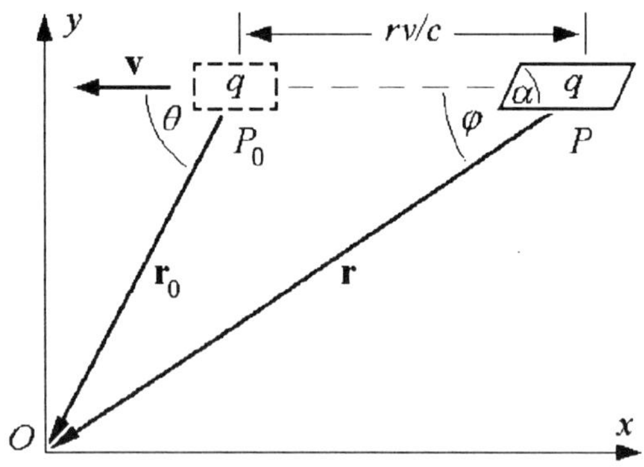

Fig. 3.4 A charge at its retarded position appears to be elongated and its vertical lines appear to be slanted.

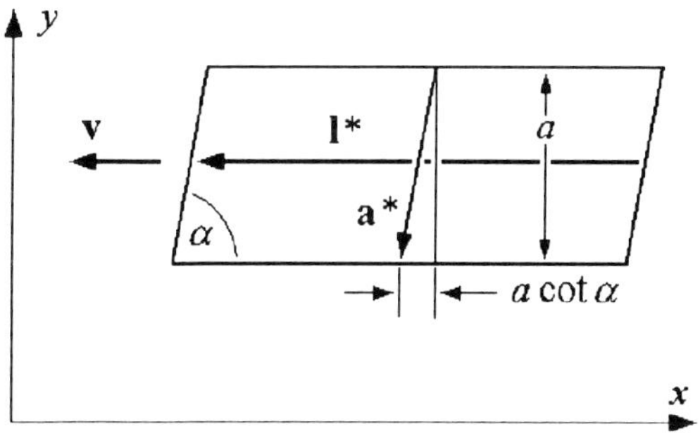

Fig. *3.5 Explanation of the vectors* **l*** *and* **a***. *The vector* **l*** *represents the retarded length of the moving charge, the vector* **a*** *represents the "slanted" thickness of the charge.*

$$\frac{dx'}{dy'} = \frac{y'v/c}{r[1-(v/c)\cos\varphi]} = \frac{y'v/c}{r[1-(\mathbf{r}\cdot\mathbf{v})r/c]} = \frac{(v/c)\sin\varphi}{1-(\mathbf{r}\cdot\mathbf{v})r/c}. \tag{3-1.12}$$

Thus, according to Eq. (3-1.12), a vertical line (x'_0 = constant, $dx'_0/dy'_0 = 0$) within the charge at the present position appears to be slanted when the charge is viewed at its retarded position (Fig. 3.4), and the angle α of the slant is given by

$$\cot\alpha = \frac{(v/c)y'}{r[1-(\mathbf{r}\cdot\mathbf{v})/rc]}. \tag{3-1.13}$$

In the derivations presented later in Chapter 4, we shall consider a moving charge in the shape of a rectangular prism of length l and thickness a. For determining the magnetic and electric fields of such a charge we shall make use of two special vectors shown in Fig. 3.5: the vector **l*** representing the retarded length of the charge, given by

$$\mathbf{l}^* = -\frac{l}{1-(\mathbf{r}\cdot\mathbf{v})/rc}\mathbf{i}, \tag{3-1.14}$$

and the vector $\mathbf{a}^*$ representing the "slanted" thickness of the charge, given by (note that $\mathbf{r} \cdot \mathbf{v} = x'v$)

$$\mathbf{a}^* = -\frac{ay'v/c}{r[1-(\mathbf{r}\cdot\mathbf{v})/rc]}\mathbf{i} - a\mathbf{j} = -\frac{ay'v/c}{r[1-(\mathbf{r}\cdot\mathbf{v})/rc]}\mathbf{i} - \frac{a(r-x'v/c)}{r[1-(\mathbf{r}\cdot\mathbf{v})/rc]}\mathbf{j}. \tag{3-1.15}$$

We shall also use the following relation derived in Example 3-1.1 for a charge moving with acceleration $\dot{\mathbf{v}} = \partial\mathbf{v}/\partial t'$

$$\nabla' \frac{1}{[r-(\mathbf{r}\cdot\mathbf{v})/c]} = \frac{\mathbf{r} - r\mathbf{v}/c + (\mathbf{r}\cdot\dot{\mathbf{v}})\mathbf{r}/c^2}{r^3[1-(\mathbf{r}\cdot\mathbf{v})/rc]^2}. \tag{3-1.16}$$

Note that if $\dot{\mathbf{v}} = 0$ (motion with constant velocity), Eq. (3-1.16) becomes

$$\nabla' \frac{1}{[r-(\mathbf{r}\cdot\mathbf{v})/c]} = \frac{\mathbf{r} - r\mathbf{v}/c}{r^3[1-(\mathbf{r}\cdot\mathbf{v})/rc]^2}. \tag{3-1.17}$$

In dealing with retarded integrals for moving electric charges, we shall frequently use the expression

$$r - (\mathbf{r}\cdot\mathbf{v})/c, \tag{3-1.18}$$

where $\mathbf{r}$ is the retarded position vector joining a retarded volume element dV' of a moving charge distribution with the point of observation. If the charge distribution moves with a constant velocity $\mathbf{v}$, this expression can be converted to the *present position* of the charge distribution, that is, to the position occupied by the volume element dV' of the charge distribution at the instant for which the electric and magnetic fields are being determined. This can be done as follows.

First, assuming that the charge distribution moves in the negative x direction and assuming that dV' is in the xy plane, we see from Fig. 3.3 that the present position vector $\mathbf{r}_0$ of dV' can be expressed in terms of the retarded position vector $\mathbf{r}$ as

$$\mathbf{r}_0 = \mathbf{r} - r\mathbf{v}/c. \tag{3-1.19}$$

Next, we write Eq. (3-1.18) as

$$\begin{aligned}[r-(\mathbf{r}\cdot\mathbf{v})/c] &= [r-x'v/c] \\ &= [(r-x'v/c)^2]^{1/2} = [r^2-2rx'v/c+x'^2v^2/c^2]^{1/2}.\end{aligned} \tag{3-1.20}$$

Adding and subtracting x'^2 and r^2v^2/c^2 to the right side of Eq. (3-1.20), we then have

$$\begin{aligned}&[r-(\mathbf{r}\cdot\mathbf{v})/c] \\ &= [r^2-2rx'v/c+x'^2v^2/c^2+x'^2-x'^2+r^2v^2/c^2-r^2v^2/c^2]^{1/2}.\end{aligned} \tag{3-1.21}$$

Let us now collect the terms on the right of Eq. (3-1.21) into three groups:

$$x'^2 - 2rx'v/c + r^2v^2/c^2, \tag{3-1.22}$$

$$r^2 - x'^2, \tag{3-1.23}$$

and

$$x'^2v^2/c^2 - r^2v^2/c^2. \tag{3-1.24}$$

By Eq. (3-1.9), the first group represents $x_0'^2$, where x_0' is the distance between the yz plane and the volume element dV' of the moving charge at its present position. The second group is simply y'^2, where y' is the (constant) y coordinate of the volume element dV'. And the third group is $-\ y'^2v^2/c^2$. We can write therefore

$$\begin{aligned}[r-(\mathbf{r}\cdot\mathbf{v})/c] &= (x_0'^2+y'^2-y'^2v^2/c^2)^{1/2} \\ &= (x_0'^2+y'^2)^{1/2}\{1-(v^2/c^2)y'^2/(x_0'^2+y'^2)\}^{1/2}.\end{aligned} \tag{3-1.25}$$

But, as can be seen from Fig. 3.3, $x_0'^2 + y'^2 = r_0^2$, and $y'^2/(x_0^2 + y'^2) = \sin^2\theta$, where θ is the angle between $\mathbf{r}_0$ and the velocity vector $\mathbf{v}$. Therefore

$$[r-(\mathbf{r}\cdot\mathbf{v})/c] = r[1-(\mathbf{r}\cdot\mathbf{v})/rc] = r_0\{1-(v^2/c^2)\sin^2\theta\}^{1/2}, \tag{3-1.26}$$

where all the quantities in the last expression are present time quantities. In obtaining Eqs. (3-1.25) and (3-1.26) we assumed that the volume element dV' of the moving charge was located in the xy plane. Clearly, however, the two equations are valid even if dV' is not in that plane, provided that we replace in these equations y'^2 by $y'^2 + z'^2$.

Expressions involving the retarded position vector $\mathbf{r}$ and its magnitude r have a very peculiar and important property which should be kept in mind when dealing with moving charges and currents. As already mentioned, a moving charge is assumed to move through different but *fixed* points of space. Therefore neither the retarded position vector $\mathbf{r}$ nor its magnitude r explicitly appearing in retarded integrals is a function of time. On the other hand, in the case of moving charges and currents, the distance r appearing in the retarded time $t' = t - r/c$ is variable and therefore is a function of time. The same applies to Eqs. (3-1.7) - (3-1.17) presented above and to all similar expressions.

▼

Example 3-1.1 Derive Eq. (3-1.16).

Let us arrange a rectangular system of coordinates so that the acceleration vector of the moving charge is in the xy plane and the velocity vector is in the negative x direction. Let the point of observation be at the origin. The position vector of the charge is then $\mathbf{r} = -x'\mathbf{i} - y'\mathbf{j}$. Using vector identity (V-7), we have

$$\nabla' \frac{1}{[r-(\mathbf{r}\cdot\mathbf{v})/c]} = -\frac{\nabla'[r-(\mathbf{r}\cdot\mathbf{v})r/c]}{[r-(\mathbf{r}\cdot\mathbf{v})/c]^2}. \qquad (3\text{-}1.27)$$

In differentiating the numerator in Eq. (3-1.27), we should remember that the numerator is retarded. However, as explained in Section 3-1, neither the position vector $\mathbf{r}$ nor its magnitude r appearing in retarded integrals is a function of time and therefore neither is affected by retardation (the charge moves through different but *fixed* points of space). The only quantity in the numerator affected by retardation is the velocity $\mathbf{v}$ which is a function of the retarded time $t - r/c$ and does change as the charge moves. Hence we can write, making use of vector identity (V-5),

$$\nabla' \frac{1}{[r-(\mathbf{r}\cdot\mathbf{v})/c]} = -\frac{\nabla' r - \nabla'[(\mathbf{r}\cdot\mathbf{v})r/c]}{[r-(\mathbf{r}\cdot\mathbf{v})/c]^2}$$
$$= -\frac{-\mathbf{r}_u + (\mathbf{r}\cdot\mathbf{v})\mathbf{r}_u/c - (r/c)\nabla'[\mathbf{r}\cdot\mathbf{v}]}{[r-(\mathbf{r}\cdot\mathbf{v})/c]^2}. \tag{3-1.28}$$

To evaluate $\nabla'[\mathbf{r}\cdot\mathbf{v}]$, we first use vector identity (V-30), obtaining

$$\nabla'[\mathbf{r}\cdot\mathbf{v}] = [\nabla'(\mathbf{r}\cdot\mathbf{v})] + \frac{\mathbf{r}_u}{c}\left[\frac{\partial(\mathbf{r}\cdot\mathbf{v})}{\partial t}\right]. \tag{3-1.29}$$

The first expression on the right can be evaluated with the help of vector identity (V-6). Note that in this expression ∇' operates upon unretarded quantities. Therefore we have

$$\nabla'(\mathbf{r}\cdot\mathbf{v}) = (\mathbf{r}\cdot\nabla')\mathbf{v} + \mathbf{r}\times(\nabla'\times\mathbf{v}) + (\mathbf{v}\cdot\nabla')\mathbf{r} + \mathbf{v}\times(\nabla'\times\mathbf{r}). \tag{3-1.30}$$

Since all the quantities in this equation are unretarded, and since the unretarded $\mathbf{v}$ does not depend on spatial coordinates, the first two terms on the right of this equation vanish. Since $\nabla'\times\mathbf{r} = 0$, the last term vanishes also. By vector identity (V-3), the remaining term is simply $-\mathbf{v}$. We thus obtain

$$\nabla'(\mathbf{r}\cdot\mathbf{v}) = -\mathbf{v}. \tag{3-1.31}$$

Taking into account that $\mathbf{r}$ in the last term of Eq. (3-1.29) is not a function of time, we have

$$\frac{\mathbf{r}_u}{c}\left[\frac{\partial(\mathbf{r}\cdot\mathbf{v})}{\partial t}\right] = \frac{\mathbf{r}_u}{c}\left[\mathbf{r}\cdot\frac{\partial\mathbf{v}}{\partial t}\right] = \frac{\mathbf{r}_u}{c}[\mathbf{r}\cdot\dot{\mathbf{v}}]. \tag{3-1.32}$$

Combining Eqs. (3-1.28), (3-1.29), (3-1.31), and (3-1.32), factoring out r in the denominator, and multiplying the numerator and the denominator by r, we finally obtain

$$\nabla' \frac{1}{[r-(\mathbf{r}\cdot\mathbf{v})/c]} = \frac{\mathbf{r} - r\mathbf{v}/c + (\mathbf{r}\cdot\dot{\mathbf{v}})\mathbf{r}/c^2}{r^3[1-(\mathbf{r}\cdot\mathbf{v})/rc]^2}. \tag{3-1.33}$$

Although all quantities in this equation refer to the retarded position of the charge, to avoid an exceedingly cumbersome notation we do not place them between the retardation brackets.

▲

3-2. Correlation Between the Electric and the Magnetic Field of a Moving Charge Distribution

There are two special cases of moving charge distributions for which there exist simple correlations between the electric and the magnetic field produced by the distributions. The first case is that of an arbitrary charge distribution moving with constant velocity. The second case is that of a point charge moving with acceleration.

Consider first a charge distribution of arbitrary size and shape moving with constant velocity **v**. Let us form the vector product of $\varepsilon_0\mathbf{v}$ and Eq. (2-2.1). Since **v** is a constant vector, we can place it under the integral sign, so that

$$\varepsilon_0\mathbf{v}\times\mathbf{E} = -\frac{1}{4\pi}\int\frac{\mathbf{v}\times\left[\nabla'\rho + \frac{1}{c^2}\frac{\partial\mathbf{J}}{\partial t}\right]}{r}dV'. \qquad (3\text{-}2.1)$$

If a charge distribution moves with constant velocity **v**, then by Eq. (3-1.4) the derivative $\partial\mathbf{J}/\partial t$ is parallel to **v**. Therefore the product $\mathbf{v}\times[\partial\mathbf{J}/\partial t]$ vanishes, and since **v** is not affected by retardation, Eq. (3-2.1) simplifies to

$$\varepsilon_0\mathbf{v}\times\mathbf{E} = -\frac{1}{4\pi}\int\frac{[\mathbf{v}\times\nabla'\rho]}{r}dV'. \qquad (3\text{-}2.2)$$

Using now the vector identity

$$\nabla'\times(\mathbf{v}\rho) = (\nabla'\times\mathbf{v})\rho - \mathbf{v}\times\nabla'\rho \qquad (3\text{-}2.3)$$

and taking into account that $\nabla' \times \mathbf{v} = 0$ and that $\mathbf{v}\rho = \mathbf{J}$, we obtain from Eq. (3-2.2)

$$\varepsilon_0 \mathbf{v} \times \mathbf{E} = \frac{1}{4\pi} \int \frac{[\nabla' \times \mathbf{J}]}{r} dV', \tag{3-2.4}$$

which, by Eq. (2-2.2), is the same as

$$\mathbf{H} = \varepsilon_0 \mathbf{v} \times \mathbf{E}. \tag{3-2.5}$$

Since $\mu_0 \mathbf{H} = \mathbf{B}$, and $\varepsilon_0 \mu_0 = 1/c^2$, this equation can also be written as

$$\mathbf{B} = (\mathbf{v} \times \mathbf{E})/c^2. \tag{3-2.6}$$

Observe that $\mathbf{E}$ in Eqs. (3-2.5) and (3-2.6) is the electric field produced by a *moving* charge distribution.

It is interesting to note that since, in the present case, the term $\partial \mathbf{J}/\partial t$ in Eq. (3-2.2) makes no contribution to $\mathbf{v} \times \mathbf{E}$, we can write Eq. (3-2.6), using Eq. (2-2.1), as

$$\mathbf{B} = -\mathbf{v} \times \frac{1}{4\pi\varepsilon_0 c^2} \int \frac{[\nabla' \rho]}{r} dV' = -\mathbf{v} \times \frac{\mu_0}{4\pi} \int \frac{[\nabla' \rho]}{r} dV', \tag{3-2.7}$$

and, assuming that the velocity is along the x axis, so that $\mathbf{v} \times \mathbf{i} = 0$, as

$$\mathbf{B} = -\mathbf{v} \times \frac{\mu_0}{4\pi} \int \frac{[(\nabla'_y + \nabla'_z)\rho]}{r} dV', \tag{3-2.8}$$

where only the components of ∇' perpendicular to $\mathbf{v}$ occur. Furthermore, using Eq. (2-2.4) and taking into account that $\partial \mathbf{J}/\partial t$ makes no contribution to $\mathbf{v} \times \mathbf{E}$ and that $\mathbf{v} \times \mathbf{i} = 0$, we can write Eq. (3-2.6) as

$$\begin{aligned}\mathbf{B} &= \mathbf{v}\times\frac{1}{4\pi\varepsilon_0 c^2}\int\left\{\frac{[\rho]}{r^3}+\frac{1}{r^2c}\left[\frac{\partial\rho}{\partial t}\right]\right\}(y\mathbf{j}+z\mathbf{k})dV' \\ &= \mathbf{v}\times\frac{\mu_0}{4\pi}\int\left\{\frac{[\rho]}{r^3}+\frac{1}{r^2c}\left[\frac{\partial\rho}{\partial t}\right]\right\}(y\mathbf{j}+z\mathbf{k})dV'.\end{aligned} \tag{3-2.9}$$

As it follows from Eqs. (3-1.7) and (3-1.8), for slowly moving charge distributions the retardation can be neglected, in which case Eq. (3-2.6) reduces to

$$\mathbf{B} = (\mathbf{v}\times\mathbf{E})/c^2, \tag{3-2.10}$$

where **E** is the ordinary electrostatic field of the charge distribution under consideration. Likewise, Eqs. (3-2.7) - (3-2.9) reduce to the corresponding equations involving unretarded charge densities.

Consider now a point charge moving with acceleration. Let us assume that the retarded position of the point charge is given by the vector **r**, and let us form the cross product of $\mathbf{r}/(r\mu_0 c)$ and Eq. (2-2.12). Since **r** for a point charge can be considered the same throughout the entire volume occupied by the charge, we can place $\mathbf{r}/r$ under the integral signs. Noting that $\mathbf{r}\times\mathbf{r} = 0$, we then obtain

$$\frac{\mathbf{r}\times\mathbf{E}}{\mu_0 cr} = \frac{1}{4\pi\varepsilon_0\mu_0 c^2}\int\left\{\frac{[\mathbf{J}]}{r^3}+\frac{1}{r^2c}\left[\frac{\partial\mathbf{J}}{\partial t}\right]\right\}\times\mathbf{r}\,dV', \tag{3-2.11}$$

and, taking into account that $\varepsilon_0\mu_0 c^2 = 1$ and using Eq. (2-2.5), we immediately obtain

$$\mathbf{H} = \frac{\mathbf{r}\times\mathbf{E}}{\mu_0 cr}, \tag{3-2.12}$$

or

$$\mathbf{B} = \frac{\mathbf{r}\times\mathbf{E}}{cr}, \tag{3-2.13}$$

where **r** is the *retarded* position vector connecting the moving point charge with the point of observation. Equations (3-2.12) and (3-2.13) show that the magnetic field of a moving point charge is always perpendicular to the electric field produced by the charge and to the radius vector joining the retarded position of the charge with the point of observation.

It is interesting to note that for a point charge moving with constant velocity, Eqs. (3-2.5) and (3-2.6) as well as Eqs. (3-2.12) and (3-2.13) hold, because Eqs. (3-2.12) and (3-2.13) are true for any acceleration, including zero acceleration. However, it is important to remember that Eqs. (3-2.12) and (3-2.13) involve the retarded position vector **r**. If the acceleration is zero, Eq. (3-2.13) reduces to Eq. (3-2.6), as is shown in Example 4-1.1.

References and Remarks for Chapter 3

1. The retarded length should not be confused with the relativistic "Lorentz-contracted length." See Section 9-1.

4

ELECTRIC AND MAGNETIC FIELDS AND POTENTIALS OF MOVING POINT AND LINE CHARGES

The finite propagation speed of electric and magnetic fields has a profound effect on the electric and magnetic fields and potentials associated with moving charge distributions. In this chapter we shall use retarded integrals for determining electric and magnetic fields and potentials of the two simplest types of moving charge distributions: a moving point charge and a moving line charge.

4-1. The Electric Field of a Uniformly Moving Point Charge[1]

Any charge distribution viewed from a sufficiently large distance constitutes a "point charge."[2] Consider a charge distribution of total charge q and density ρ confined to a small rectangular prism (Fig. 4.1) whose center is located at the point x', y' in the xy plane of a rectangular system of coordinates, and whose sides l, a, and b are parallel to the x, y, and z axis, respectively. Let the point of observation be at the origin of the coordinates, and let the distance between the center of the prism and the origin be $r_0 \gg a, b, l$. Viewed from the origin, this

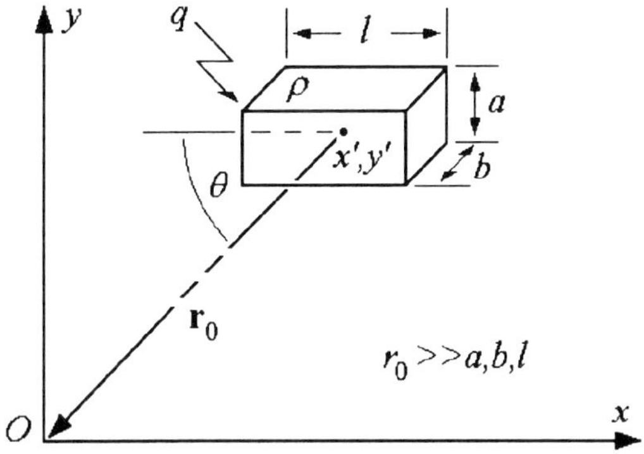

Fig. 4.1 A charge of uniform density ρ is confined to a small rectangular prism. The total charge of the prism is q. The charge constitutes a point charge when viewed from a distance large compared to the linear dimensions of the prism.

charge distribution constitutes a point charge.[3] Let the charge move with uniform velocity $\mathbf{v} = -v\mathbf{i}$. We want to find the electric and magnetic fields of this charge at the point of observation.

To find the electric field produced by this charge, we shall use Eq. (2-2.1). First we eliminate from Eq. (2-2.1) the term with the current density **J**. We can do so with the help of Eq. (3-1.4). Since the velocity of our charge is $\mathbf{v} = v_x\mathbf{i} = -v\mathbf{i}$, and since the charge moves without acceleration so that $\dot{\mathbf{v}} = 0$, Eq. (3-1.4) gives

$$\frac{\partial \mathbf{J}}{\partial t} = -\left(v_x \frac{\partial \rho}{\partial x'}\right)\mathbf{v} = -v^2 \frac{\partial \rho}{\partial x'}\mathbf{i}. \qquad (4\text{-}1.1)$$

Substituting Eq. (4-1.1) into Eq. (2-2.1), we then have for the electric field of the charge

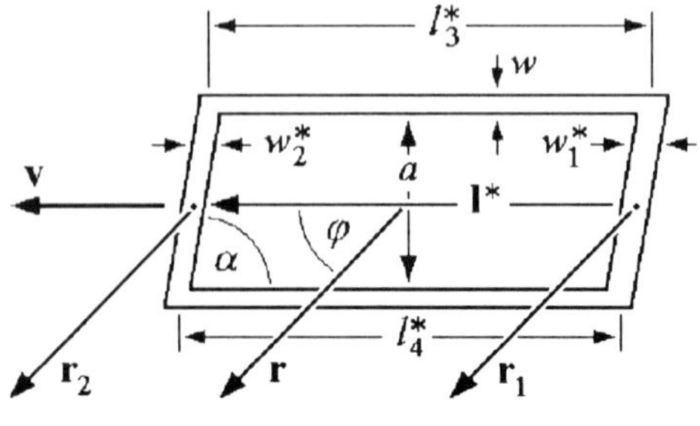

Fig. 4.2 When the charge shown in Fig. 4.1 is moving and is at a retarded position, its apparent length, shape, and thickness of its front and back surface layers are no longer the same as for the stationary charge. (All **r***'s meet at the origin).*

$$\mathbf{E} = -\frac{1}{4\pi\varepsilon_0}\int \frac{\left[\nabla'\rho - \dfrac{v^2}{c^2}\dfrac{\partial\rho}{\partial x'}\mathbf{i}\right]}{r}dV'. \tag{4-1.2}$$

Observe that in this equation ∇' and $\partial/\partial x'$ operate on the *unretarded* ρ, so that in computing $\nabla'\rho$ and $\partial\rho/\partial x'$ we must use the ordinary, unretarded, shape and size of the prism. Since ρ is constant within the prism, $\nabla'\rho = 0$ within it, and the only contribution to $\nabla'\rho$ comes from the surface layer of the prism, where ρ changes from ρ (inside the prism) to 0 (outside the prism). Let the thickness of the surface layer be w. Taking into account that $\nabla'\rho$ represents the rate of change of ρ in the direction of the greatest rate of change, we then have $\nabla'\rho = (\rho/w)\mathbf{n}_{in}$, where $\mathbf{n}_{in}$ is a unit vector normal to the surface layer and pointing *into* the prism. Hence $\nabla'\rho$ for the right, left, top, bottom, front, and back surfaces of the charge (prism) are $-\rho/w\mathbf{i}$, $\rho/w\mathbf{i}$, $-\rho/w\mathbf{j}$, $\rho/w\mathbf{j}$, $-\rho/w\mathbf{k}$, and $\rho/w\mathbf{k}$, respectively. Likewise, $\partial\rho/\partial x'$ is zero in the interior of the charge and is different from zero only in the left and in the right surface layers of the charge, where

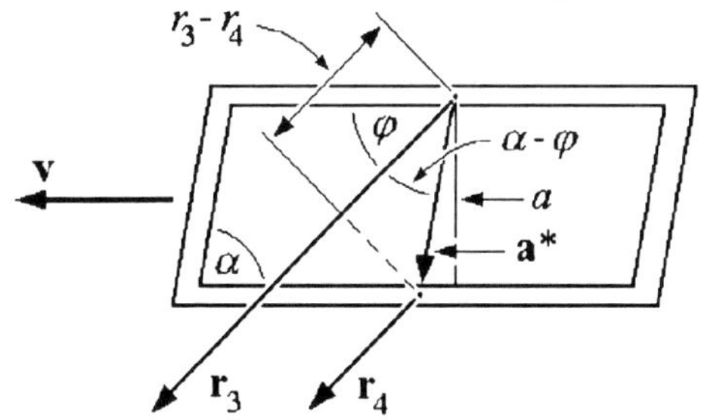

Fig. 4.3 The relations between $\mathbf{r}_3$, $\mathbf{r}_4$, *and* $\mathbf{a}^*$ *for the moving charge at a retarded position. (The two* $\mathbf{r}$*'s meet at the origin.)*

$\partial\rho/\partial x' = \rho/w$ in the left surface layer and $\partial\rho/\partial x' = -\rho/w$ in the right surface layer.

The volume integral of Eq. (4-1.2) can be split therefore into six integrals, one over each of the six surface layers corresponding to the six surfaces of the charge (prism). However, since the center of the charge is in the xy plane ($z' = 0$), the integrals over the two surface layers parallel to the xy plane cancel each other, because $\nabla'\rho$ for one of the layers is opposite to that for the other layer, while r is the same for both layers. Thus only the four integrals over the layers parallel to the xz and yz planes remain. Let us designate the retarded distances from these layers to the point of observation as r_1, r_2, r_3, and r_4 (see Figs. 4.2 and 4.3). Since the linear dimensions of the charge are much smaller than r_1, r_2, r_3, and r_4, we can replace each integral over a surface layer by the product of the integrand and the volume of the corresponding layer. However, the integration in Eq. (4-1.2) is over the *effective* (retarded) volume of the charge, and therefore we must use not the true volume of the surface layers, but their effective volume. The effective volume of the surface layers is not the same as their actual volume, because, in accordance with Eq. (3-1.7), the length l of the two layers parallel to the xz plane must be replaced by

$$l^* = \frac{l}{1-(\mathbf{r}\cdot\mathbf{v})/rc}, \tag{4-1.3}$$

and because, also in accordance with Eq. (3-1.7), the thickness w of the two layers parallel to the yz plane must be replaced by

$$w^* = \frac{w}{1-(\mathbf{r}\cdot\mathbf{v})/rc}. \tag{4-1.4}$$

Equation (4-1.2) becomes therefore

$$\mathbf{E} = -\frac{1}{4\pi\varepsilon_0}\left[\frac{\rho/w}{r_1}abw_1^*(-\mathbf{i}) + \frac{\rho/w}{r_2}abw_2^*\mathbf{i} + \frac{\rho/w}{r_3}bl_3^*w(-\mathbf{j}) + \frac{\rho/w}{r_4}bl_4^*w\mathbf{j} + \left(\frac{v^2}{c^2}\right)\left(\frac{\rho/w}{r_1}abw_1^*\mathbf{i} + \frac{\rho/w}{r_2}abw_2^*(-\mathbf{i})\right)\right] \tag{4-1.5}$$

or, substituting l^* and w^* from Eqs. (4-1.3) and (4-1.4),

$$\begin{aligned}\mathbf{E} = -\frac{1}{4\pi\varepsilon_0}\Bigg[&\frac{\rho/w}{r_1-\mathbf{r}_1\cdot\mathbf{v}/c}abw(-\mathbf{i}) + \frac{\rho/w}{r_2-\mathbf{r}_2\cdot\mathbf{v}/c}abw\mathbf{i}\\ &+ \frac{\rho/w}{r_3-\mathbf{r}_3\cdot\mathbf{v}/c}blw(-\mathbf{j}) + \frac{\rho/w}{r_4-\mathbf{r}_4\cdot\mathbf{v}/c}blw\mathbf{j}\\ &+ \left(\frac{v^2}{c^2}\right)\left(\frac{\rho/w}{r_1-\mathbf{r}_1\cdot\mathbf{v}/c}abw\mathbf{i} + \frac{\rho/w}{r_2-\mathbf{r}_2\cdot\mathbf{v}/c}abw(-\mathbf{i})\right)\Bigg],\end{aligned} \tag{4-1.6}$$

which simplifies to

$$\begin{aligned}\mathbf{E} = -\frac{\rho b}{4\pi\varepsilon_0}\Bigg[&\left(1-\frac{v^2}{c^2}\right)\left(\frac{1}{r_2-\mathbf{r}_2\cdot\mathbf{v}/c} - \frac{1}{r_1-\mathbf{r}_1\cdot\mathbf{v}/c}\right)a\mathbf{i}\\ &+ \left(\frac{1}{r_4-\mathbf{r}_4\cdot\mathbf{v}/c} - \frac{1}{r_3-\mathbf{r}_3\cdot\mathbf{v}/c}\right)l\mathbf{j}\Bigg].\end{aligned} \tag{4-1.7}$$

As can be seen from Figs. 4.2 and 4.3, the differences of the fractions in these equations are simply the increments of the function $1/(r - \mathbf{r}\cdot\mathbf{v}/c)$ associated with the displacement of the tail of $\mathbf{r}$ over the distances represented by the vector $\mathbf{l}^*$ [in the $\mathbf{i}$ component of Eq. (4-1.7)] and by the vector $\mathbf{a}^*$ [in the $\mathbf{j}$ component of Eq. (4-1.7)]. Therefore we can write Eq. (4-1.7) as[4]

$$\mathbf{E} = -\frac{\rho b}{4\pi\varepsilon_0}\left\{\left(1-\frac{v^2}{c^2}\right)\left[\left(\nabla'\frac{1}{r-\mathbf{r}\cdot\mathbf{v}/c}\right)\cdot\mathbf{l}^*\right]a\mathbf{i} + \left[\left(\nabla'\frac{1}{r-\mathbf{r}\cdot\mathbf{v}/c}\right)\cdot\mathbf{a}^*\right]l\mathbf{j}\right\}. \tag{4-1.8}$$

Substituting the gradient from Eq. (3-1.17), remembering that $\dot{\mathbf{v}} = 0$, and substituting $\mathbf{l}^*$ and $\mathbf{a}^*$ from Eqs. (3-1.14) and (3-1.15), we have

$$\begin{aligned}\mathbf{E} = \frac{\rho b}{4\pi\varepsilon_0}\Bigg[&\left(1-\frac{v^2}{c^2}\right)\left(\frac{\mathbf{r}-r\mathbf{v}/c}{r^3(1-\mathbf{r}\cdot\mathbf{v}/rc)^2}\cdot\mathbf{i}\right)\frac{la}{1-\mathbf{r}\cdot\mathbf{v}/rc}\mathbf{i}\\ &+\left(\frac{\mathbf{r}-r\mathbf{v}/c}{r^3(1-\mathbf{r}\cdot\mathbf{v}/rc)^2}\cdot\mathbf{i}\right)\frac{y'v/c}{r(1-\mathbf{r}\cdot\mathbf{v}/rc)}al\mathbf{j}\\ &+\left(\frac{\mathbf{r}-r\mathbf{v}/c}{r^3(1-\mathbf{r}\cdot\mathbf{v}/rc)^2}\cdot\mathbf{j}\right)\frac{r-x'v/c}{r(1-\mathbf{r}\cdot\mathbf{v}/rc)}al\mathbf{j}\Bigg].\end{aligned} \tag{4-1.9}$$

Simplifying and taking into account that $\mathbf{r}\cdot\mathbf{i} = -x'$, $\mathbf{r}\cdot\mathbf{j} = -y'$, $\mathbf{v}\cdot\mathbf{i} = -v$, $\mathbf{v}\cdot\mathbf{j} = 0$, and $\mathbf{r}\cdot\mathbf{v} = x'v$, we obtain

$$\begin{aligned}\mathbf{E} &= \frac{\rho abl}{4\pi\varepsilon_0 r^3[1-\mathbf{r}\cdot\mathbf{v}/rc]^3}\left[\left(1-\frac{v^2}{c^2}\right)(-x'+rv/c)\,\mathbf{i} + (-x'+rv/c)\frac{y'v/c}{r}\mathbf{j} + (-y')\frac{r-x'v/c}{r}\mathbf{j}\right]\\ &= \frac{\rho abl}{4\pi\varepsilon_0 r^3[1-\mathbf{r}\cdot\mathbf{v}/rc]^3}\left[\left(1-\frac{v^2}{c^2}\right)(-x'\mathbf{i}-r\mathbf{v}/c) + \left(1-\frac{v^2}{c^2}\right)(-y)\mathbf{j}\right],\end{aligned} \tag{4-1.10}$$

and finally, noting that $\mathbf{r} = -x'\mathbf{i} - y'\mathbf{j}$, and that $\rho abl = q$,

$$\mathbf{E} = \frac{q(1-v^2/c^2)}{4\pi\varepsilon_0 r^3[1-\mathbf{r}\cdot\mathbf{v}/rc]^3}\left[\mathbf{r}-\frac{r\mathbf{v}}{c}\right]. \tag{4-1.11}$$

Equation (4-1.11) expresses **E** in terms of the *retarded* position of the charge specified by the retarded position vector **r** (see Fig. 3.4). Usually it is desirable to express **E** in terms of the *present* position of the charge specified by the present position vector $\mathbf{r}_0$ (see Fig. 3.4). We can convert Eq. (4-1.11) from **r** to $\mathbf{r}_0$ by using Eqs. (3-1.19) and (3-1.27). According to Eq. (3-1.19),

$$\mathbf{r} - r\mathbf{v}/c = \mathbf{r}_0, \tag{4-1.12}$$

so that the last factor in Eq. (4-1.11) is simply the present position vector $\mathbf{r}_0$. Substituting Eq. (4-1.12) and Eq. (3-1.27) into Eq. (4-1.11), we obtain the desired equation for the electric field of a uniformly moving point charge expressed in terms of the present position of the charge

$$\mathbf{E} = \frac{q(1 - v^2/c^2)}{4\pi\varepsilon_0 r_0^3[1 - (v^2/c^2)\sin^2\theta]^{3/2}}\mathbf{r}_0. \tag{4-1.13}$$

This equation (in a different notation) was first derived by Oliver Heaviside in 1888 on the basis of Maxwell's equations by using the "operational calculus" that he invented.[5]

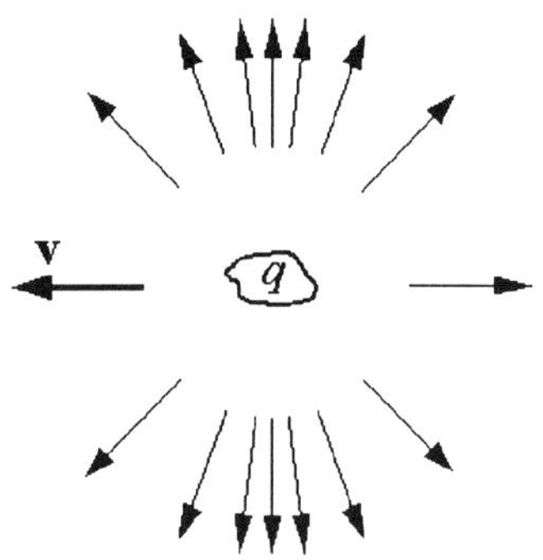

Fig. 4.4 As was first noticed by Heaviside, the electric field of a moving point charge concentrates itself in the direction perpendicular to the direction of motion of the charge and decreases along the line of the motion.

There are two interesting properties of Eq. (4-1.13). First, as was noted by Heaviside, with increasing velocity of the charge the electric field of the charge concentrates itself more and more about the equatorial plane, $\theta = \pi/2$, and decreases along the line of motion, $\theta = 0$. This effect is shown in Fig. 4.4. Second, the electric field *appears to originate* at the charge in its present position. This, of course, is merely an illusion, because by supposition the distance between the charge and the point of observation is much greater than the linear dimensions of the charge, so that neither Eq. (4-1.11) nor Eq. (4-1.13) gives us any information concerning the structure of the field close to the charge. Note also that because of the finite speed of the propagation of the field signals and light signals one can never *observe* the charge at its present position. In fact, the charge could have stopped after sending the field signal from its retarded position, and even then Eq. (4-1.13) would remain valid, although in this case Eq. (4-1.13) would apply to the "projected," or "anticipated," present position of the charge.

▼

Example 4-1.1 Show that for a point charge moving without acceleration Eq. (3-2.13) reduces to (3-2.6).

According to Eq. (4-1.12), the retarded position vector of the charge can be expressed in terms of the present position as

$$\mathbf{r} = \mathbf{r}_0 + r\mathbf{v}/c. \tag{4-1.14}$$

Substituting Eq. (4-1.14) into Eq. (3-2.13), we have

$$\mathbf{B} = \frac{\mathbf{r} \times \mathbf{E}}{cr} = \frac{(\mathbf{r}_0 + r\mathbf{v}/c) \times \mathbf{E}}{cr} = \frac{\mathbf{r}_0 \times \mathbf{E}}{cr} + \frac{(r\mathbf{v}/c) \times \mathbf{E}}{cr}. \tag{4-1.15}$$

Since, by Eq. (4-1.13), $\mathbf{E}$ is directed along $\mathbf{r}_0$, $\mathbf{r}_0 \times \mathbf{E} = 0$, and we are left with

$$\mathbf{B} = (\mathbf{v} \times \mathbf{E})/c^2, \tag{4-1.16}$$

which was to be proved.

Example 4-1.2 Equation (4-1.13) represents a "snapshot" of the electric field of a moving point charge, since it does not express the field as a function of time. Modify Eq. (4-1.13) so that it shows how the field changes as the charge moves.

Let us assume that the "snapshot" is for $t = 0$. If the charge moves in the $-x$ direction, the functional dependence of $\mathbf{E}$ on the x coordinate will be preserved for $t \neq 0$ if we express Eq. (4-1.13) in terms of x_0' and replace x_0' by $x_0' - vt$. From Eqs. (3-1.26) and (3-1.27), we have

$$\begin{aligned} r_0\{1-(v^2/c^2)\sin^2\theta\}^{1/2} &= (x_0'^2+y'^2-y'^2v^2/c^2)^{1/2} \\ &= [x_0'^2+(1-v^2/c^2)y'^2]^{1/2}. \end{aligned} \tag{4-1.17}$$

Replacing in Eq. (4-1.17) x_0' by $x_0' - vt$, we obtain

$$r_0\{1-(v^2/c^2)\sin^2\theta\}^{1/2} = [(x_0'-vt)^2+(1-v^2/c^2)y'^2]^{1/2}, \tag{4-1.18}$$

where x_0' is now the x coordinate of the point charge at $t = 0$. Expressing $\mathbf{r}_0$ in terms of its components and replacing x_0' by $x_0' - vt$, we similarly have $\mathbf{r}_0 = (x_0' - vt)\mathbf{i} + y'\mathbf{j}$. Therefore Eq. (4-1.13) can be written as

$$\mathbf{E} = \frac{q(1-v^2/c^2)\{(x_0'-vt)\mathbf{i}+y'\mathbf{j}\}}{4\pi\varepsilon_0\{(x_0'-vt)^2+(1-v^2/c^2)y'^2\}^{3/2}}, \tag{4-1.19}$$

where the dependence of $\mathbf{E}$ on t is shown explicitly. This equation holds for the charge moving parallel to the x axis in the xy plane. If it moves parallel to the x axis anywhere in space, y'^2 in this equation should be replaced by $(y'^2 + z'^2)$. ▲

4-2. The Magnetic Field of a Uniformly Moving Point Charge

Although by using Eq. (2-2.2) or Eq. (2-2.5), we can find the magnetic field of a uniformly moving point charge in the same manner as we found the electric field in Section 4-1 (see Example 4-2.1), it is much easier to find it from the known electric field by using Eq. (3-2.5) or Eq. (3-2.6).

Applying Eq. (3-2.5) to Eq. (4-1.11), we obtain for the magnetic field in terms of the retarded position of the charge

$$\mathbf{H} = \frac{q[1 - v^2/c^2]}{4\pi r^3[1 - \mathbf{r} \cdot \mathbf{v}/rc]^3}[\mathbf{v} \times \mathbf{r}]. \tag{4-2.1}$$

Applying Eq. (3-2.5) to Eq. (4-1.13), we obtain for the magnetic field in terms of the present position of the charge

$$\mathbf{H} = \frac{q(1 - v^2/c^2)}{4\pi r_0^3\{1 - (v^2/c^2)\sin^2\theta\}^{3/2}}[\mathbf{v} \times \mathbf{r}_0]. \tag{4-2.2}$$

▼

Example 4-2.1 Find the magnetic field of a uniformly moving point charge shown in Fig. 4.1 by using Eq. (2-2.2),

$$\mathbf{H} = \frac{1}{4\pi}\int \frac{[\nabla' \times \mathbf{J}]}{r} dV'. \tag{4-2.3}$$

To use Eq. (4-2.3), we need to know $\nabla' \times \mathbf{J}$ associated with the charge under consideration. The moving charge constitutes a current density $\mathbf{J} = \rho\mathbf{v}$. Since $\mathbf{v}$ is not a function of x', y', z', we have $\nabla' \times \mathbf{J} = \nabla'\rho \times \mathbf{v}$. But ρ is constant within the charge, and therefore the only contribution to $\nabla' \times \mathbf{J}$ comes from the surface layer of the charge, where ρ changes from ρ (inside the charge) to 0 (outside the charge). Using the values for $\nabla'\rho$ obtained in Section 4-1, we then have for $\nabla' \times \mathbf{J}$ of the top, bottom, front, and back surface layers of the charge (prism) $-\rho v/w\mathbf{k}$, $\rho v/w\mathbf{k}$, $\rho v/w\mathbf{j}$, and $-\rho/w\mathbf{j}$, respectively; the left and right surface layers make no contribution to $\nabla' \times \mathbf{J}$, because $\mathbf{v}$ and $\nabla'\rho$ are parallel (or antiparallel) there. Furthermore, since $\nabla' \times \mathbf{J}$ in the front surface layer is opposite to $\nabla' \times \mathbf{J}$ in the back surface layer, while both surface layers are at the same distance r from the point of observation, the contributions of these two layers to the integral in

Eq. (4-2.3) cancel each other, so that only the top and the bottom surface layers contribute to the magnetic field of the charge.

Since the linear dimensions of the charge are much smaller than r_3 and r_4, (see Fig. 4.3), we can replace the integrals over the two surface layers by the product of the integrand and the volumes of the corresponding layers. Using Eq. (4-2.3) and taking into account the effective volume of the boundary layers (see Sections 3-1 and 4-1), we have, as in Eqs. (4-1.5)-(4-1.7),

$$\begin{aligned}\mathbf{H} &= -\frac{1}{4\pi}\left[\frac{\rho v/w}{r_3 - \mathbf{r}_3 \cdot \mathbf{v}/c} wbl\mathbf{k} + \frac{\rho v/w}{r_4 - \mathbf{r}_4 \cdot \mathbf{v}/c} wbl\mathbf{k}\right] \\ &= -\frac{\rho vbl}{4\pi}\left[\frac{1}{r_3 - \mathbf{r}_3 \cdot \mathbf{v}/c} - \frac{1}{r_4 - \mathbf{r}_4 \cdot \mathbf{v}/c}\right]\mathbf{k}.\end{aligned} \tag{4-2.4}$$

The difference of the two fractions in the last expression is simply the increment of the function $1/(r - \mathbf{r} \cdot \mathbf{v}/c)$ associated with the displacement of the tail of $\mathbf{r}$ over the distance represented by the vector $\mathbf{a}^*$ (see Fig 4.3). Therefore, using Eqs. (3-1.17) and (3-1.15), we can write Eq. (4-2.4) as

$$\begin{aligned}\mathbf{H} = -\frac{\rho bvl}{4\pi}&\left[\left(\frac{\mathbf{r} - r\mathbf{v}/c}{r^3(1 - \mathbf{r} \cdot \mathbf{v}/rc)^2} \cdot \mathbf{i}\right)\frac{y'v/c}{r(1 - \mathbf{r} \cdot \mathbf{v}/rc)} a\right. \\ &+ \left.\left(\frac{\mathbf{r} - r\mathbf{v}/c}{r^3(1 - \mathbf{r} \cdot \mathbf{v}/rc)^2} \cdot \mathbf{j}\right)\frac{r - x'v/c}{r(1 - \mathbf{r} \cdot \mathbf{v}/rc)} a\right]\mathbf{k}.\end{aligned} \tag{4-2.5}$$

Simplifying and taking into account that $\mathbf{r} \cdot \mathbf{i} = -x'$, $\mathbf{r} \cdot \mathbf{j} = -y'$, $\mathbf{v} \cdot \mathbf{i} = -v$, $\mathbf{v} \cdot \mathbf{j} = 0$, and $\mathbf{r} \cdot \mathbf{v} = x'v$, we obtain

$$\begin{aligned}\mathbf{H} &= -\frac{qv}{4\pi r^3[1 - \mathbf{r} \cdot \mathbf{v}/rc]^3}[(-x' + rv/c)y'v/rc + (-y')(1 - x'v/rc)]\mathbf{k} \\ &= \frac{qv[1 - v^2/c^2]y}{4\pi r^3[1 - \mathbf{r} \cdot \mathbf{v}/rc]^3} y\mathbf{k},\end{aligned} \tag{4-2.6}$$

which, noting that $vy'\mathbf{k} = \mathbf{v} \times \mathbf{r}$, is the same as Eq. (4-2.1).[6]

▲

4-3. The Electric and Magnetic Fields of a Line Charge Uniformly Moving Along its Length

Consider a line charge of finite length L, cross-sectional area S, charge density ρ, and linear charge density $\lambda = \rho S$ moving with constant velocity $\mathbf{v}$ parallel to the x axis of a rectangular system of coordinates in the negative direction of the axis and at a distance R above the axis (Fig. 4.5). Let the point of observation O be at the origin. What is the electric field at O at the time t when the leading end of the charge is at a distance L_2 from the y axis?

We can find the electric field of the moving charge by using Eq. (2-2.1) or Eq. (2-2.4) if we know the retarded position of the

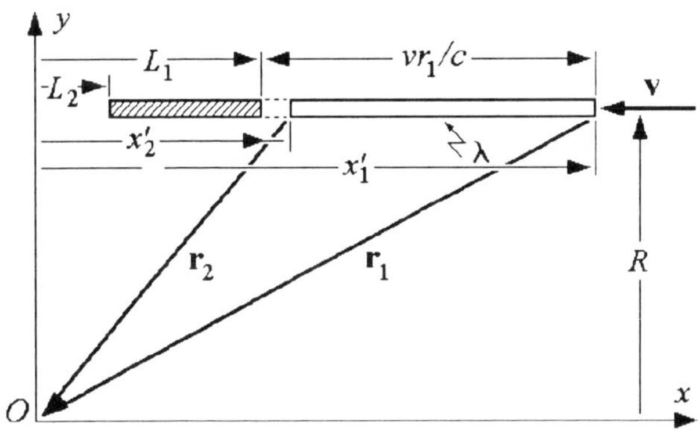

Fig. 4.5 A line charge of linear density λ is moving with constant velocity $\mathbf{v}$. *The retarded positions of the trailing and leading ends of the charge are* x_1' and x_2', *respectively. The present positions of the two ends are* L_1 *and* L_2, *respectively. The distance between the trajectory of the charge and the x axis is R. The point of observation O is at the origin. The "retarded," or "effective," length of the charge is longer than its true length.*

charge corresponding to the time for which the field is computed. We can determine this position as follows.

First, let us determine the retarded position x_2' of the leading end of the charge corresponding to the time t, that is, the position from which the leading end sends out its field signal which arrives at O at the time t. If the retarded distance between O and the leading end is r_2, then the time it takes for the signal to travel from the leading end to O is r_2/c. During this time the charge travels a distance $v(r_2/c)$. Therefore at the moment when the leading end sends out its field signal, the position of the leading end is

$$x_2' = L_2 + r_2 v/c. \tag{4-3.1}$$

Next, let us find the retarded position x_1' of the trailing end of the charge corresponding to the time t. If the retarded distance between O and the trailing end is r_1, then the time it takes for the signal to travel from the trailing end to O is r_1/c. During this time the charge travels a distance $v(r_1/c)$. Hence, at the moment when the trailing end sends out its signal, the position of the trailing end is

$$x_1' = L_1 + r_1 v/c. \tag{4-3.2}$$

The x component of the electric field. We are now ready to find the electric field of the charge by using Eq. (2-2.1) or Eq. (2-2.4). The easiest way to find the x component of the electric field of the charge under consideration is to use Eq. (2-2.1). According to this equation, the x component of the field is due to the x components of $[\nabla'\rho]$ and $[\partial\mathbf{J}/\partial t]$ of the moving charge. For the line charge under consideration, these components exist only at the leading and trailing ends of the charge and are the same as for the moving charged prism discussed in the preceding sections of this chapter: $[\nabla'\rho]_x = \rho/w$ for the leading end, and $[\nabla'\rho]_x = -\rho/w$ for the trailing end, $[\partial\mathbf{J}/\partial t]_x = -v^2\rho/w$ for the leading end, and $[\partial\mathbf{J}/\partial t]_x = v^2\rho/w$ for the trailing end, where w is the thickness of the surface layer of the charge (this is the actual thickness, not the retarded one). Since the surface layer of the charge may be assumed as thin as one wishes, the retarded volume integral in Eq. (2-2.1), as far as the x component of the field is concerned, reduces to the product of the integrand and the volume of the surface layers of the leading and trailing ends of the charge at their retarded positions. By Eq. (4-1.4), for the leading end, this volume is, using the asterisk to indicate values evaluated at retarded positions,

$$w_2^* S = \frac{wS}{1-(\mathbf{r}_2 \cdot \mathbf{v})/r_2 c}, \tag{4-3.3}$$

and for the trailing end it is

$$w_1^* S = \frac{wS}{1-(\mathbf{r}_1 \cdot \mathbf{v})/r_1 c} \, . \tag{4-3.4}$$

The x component of the electric field is therefore

$$E_x = -\frac{\rho S(1-v^2/c^2)}{4\pi\varepsilon_0}\left(\frac{1}{r_2[1-(\mathbf{r}_2\cdot\mathbf{v})/r_2c]} - \frac{1}{r_1[1-(\mathbf{r}_1\cdot\mathbf{v})/r_1c]}\right), \tag{4-3.5}$$

or

$$E_x = -\frac{\lambda(1-v^2/c^2)}{4\pi\varepsilon_0}\left(\frac{1}{r_2 - x_2' v/c} - \frac{1}{r_1 - x_1' v/c}\right). \tag{4-3.6}$$

Equation (4-3.6) gives the electric field in terms of the retarded position of the charge. We shall now convert it to the present position of the charge (that is, the actual position of the charge at the time t). The calculations are similar to those used for deriving Eqs. (3-1.20)-(3-1.27). First, we note that, by Eq. (4-3.1),

$$L_2^2 = x_2'^2 - 2x_2 r_2 v/c + r_2^2 v^2/c^2. \tag{4-3.7}$$

Next, we write the denominator of the first fraction inside the parentheses of Eq. (4-3.6) as

$$r_2 - x_2' v/c = [(r_2 - x_2' v/c)^2]^{1/2} = (r_2^2 - 2r_2 x_2' v/c + x_2'^2 v^2/c^2)^{1/2}. \quad (4\text{-}3.8)$$

Adding and subtracting x'^2 and $r_2^2v^2/c^2$ to the right side of Eq. (4-3.8), we then have

$$r_2 - x_2' v/c \quad (4\text{-}3.9)$$
$$= (r_2^2 - 2r_2 x_2' v/c + x_2'^2 v^2/c^2 + x_2'^2 - x_2'^2 + r_2^2 v^2/c^2 - r_2^2 v^2/c^2)^{1/2}.$$

Let us now collect the terms on the right of Eq. (4-3.9) into three groups:

$$x_2'^2 - 2r_2 x_2' v/c + r_2^2 v^2/c^2, \quad (4\text{-}3.10)$$

$$r_2^2 - x_2'^2, \quad (4\text{-}3.11)$$

and

$$x_2'^2 v^2/c^2 - r_2^2 v^2/c^2. \quad (4\text{-}3.12)$$

By Eq. (4-3.7), the first group represents L_2^2. The second group is simply R^2 (see Fig. 4.5). And the third group is $-R^2v^2/c^2$.

Similar relations hold for the denominator of the second fraction inside the parentheses of Eq. (4-3.6). Therefore Eq. (4-3.6) transforms to

$$E_x = \frac{\lambda(1 - v^2/c^2)}{4\pi\varepsilon_0 R}\left[\frac{1}{(L_1^2/R^2 + 1 - v^2/c^2)^{1/2}} - \frac{1}{(L_2^2/R^2 + 1 - v^2/c^2)^{1/2}}\right], \quad (4\text{-}3.13)$$

where only the present time quantities appear.

The y component of the electric field. The easiest way to find the y component of the electric field of the charge under consideration is to use Eq. (2-2.4). Only the first integral of Eq. (2-2.4) makes a contribution to the y component of the field, because $\partial\mathbf{J}/\partial t$ has no y component. Separating this integral into two integrals, we then have

$$E_y = -\frac{1}{4\pi\varepsilon_0}\int\frac{[\rho]}{r^3}RdV' - \frac{1}{4\pi\varepsilon_0}\int\frac{1}{r^2c}\left[\frac{\partial\rho}{\partial t}\right]RdV'. \quad (4\text{-}3.14)$$

The first integral in Eq. (4-3.14) is the same as for a stationary charge, except that the integration must be extended over the retarded (effective) length of the charge. Designating the contribution of the first integral as E_{1y} and noting that $r = (x'^2 + R^2)^{1/2}$, we obtain

$$E_{1y} = -\frac{1}{4\pi\varepsilon_0}\int\frac{\rho}{r^3}RdV' = -\frac{\rho S}{4\pi\varepsilon_0}\int_{x_2'}^{x_1'}\frac{R}{(x'^2+R^2)^{3/2}}dx', \quad (4\text{-}3.15)$$

or

$$E_{1y} = -\frac{\lambda}{4\pi\varepsilon_0 R}\left[\frac{x_1'}{(x_1'^2+R^2)^{1/2}} - \frac{x_2'}{(x_2'^2+R^2)^{1/2}}\right] = \frac{\lambda}{4\pi\varepsilon_0 R}\left(\frac{x_2'}{r_2} - \frac{x_1'}{r_1}\right). \quad (4\text{-}3.16)$$

In order to evaluate the contribution of E_{2y} of the second integral of Eq. (4-3.14) to the total field, we must determine the value of the derivative $[\partial\rho/\partial t]$. According to the notation convention for retarded quantities explained in Chapter 1, this derivative is the ordinary derivative $\partial\rho/\partial t$ used at the retarded position of the moving charge. By Eq. (3-1.3), taking into account that for our charge $\mathbf{v} = -v\mathbf{i}$, $[\partial\rho/\partial t]$ is then simply $v\partial\rho/\partial x'$. Since ρ is constant within the line charge, only the leading and the trailing ends of the charge contribute to this expression, and the contributions are $v\rho/w$ and $-v\rho/w$, respectively. The electric field E_{2y} is therefore

$$E_{2y} = -\frac{R}{4\pi\varepsilon_0 c}\int\frac{v\rho/w}{r_2^2}dV_2' + \frac{R}{4\pi\varepsilon_0 c}\int\frac{v\rho/w}{r_1^2}dV_1', \quad (4\text{-}3.17)$$

where the integration is over the surface layers of the leading and trailing ends of the charge at the retarded positions of the charge. Since the thickness of the surface layers is much smaller than r_1 and r_2, we can replace the integrals, as before for E_x, by the products of the integrands and the volumes of integration (the volumes of the respective surface layers). Using the relations $dV_2' = w_2{*}S$, $dV_1' = w_1{*}S$, and using Eqs. (4-3.3) and (4-3.4) , we then have

$$E_{2y} = -\frac{R}{4\pi\varepsilon_0 c}\left[\frac{v\rho/w}{r_2^2 - r_2(\mathbf{r}_2\cdot\mathbf{v})/c}wS + \frac{v\rho/w}{r_1^2 - r_1(\mathbf{r}_1\cdot\mathbf{v})/c}wS\right]$$
$$= \frac{\lambda v R}{4\pi\varepsilon_0 c}\left[\frac{1}{r_1(r_1 - x_1'v/c)} - \frac{1}{r_2(r_2 - x_2'v/c)}\right]. \tag{4-3.18}$$

Adding Eqs. (4-3.16) and (4-3.18), we obtain for the y component of the field

$$E_y = \frac{\lambda}{4\pi\varepsilon_0 R}\left[-\frac{x_1'}{r_1} + \frac{R^2 v/c}{r_1(r_1 - x_1'v/c)} + \frac{x_2'}{r_2} - \frac{R^2 v/c}{r_2(r_2 - x_2'v/c)}\right]$$
$$= \frac{\lambda}{4\pi\varepsilon_0 R}\left[\frac{x_2'(r_2 - x_2'v/c) - R^2 v/c}{r_2(r_2 - x_2'v/c)} - \frac{x_1'(r_1 - x_1'v/c) - R^2 v/c}{r_1(r_1 - x_1'v/c)}\right], \tag{4-3.19}$$

or

$$E_y = \frac{\lambda}{4\pi\varepsilon_0 R}\left[\frac{x_2'r_2 - x_2'^2 v/c - R^2 v/c}{r_2(r_2 - x_2'v/c)} - \frac{x_1'r_1 - x_1'^2 v/c - R^2 v/c}{r_1(r_1 - x_1'v/c)}\right]. \tag{4-3.20}$$

But $x_1'^2 v/c + R^2 v/c = r_1^2 v/c$ and $x_2'^2 v/c + R^2 v/c = r_2^2 v/c$. Therefore

$$E_y = \frac{\lambda}{4\pi\varepsilon_0 R}\left(\frac{x_2' - r_2 v/c}{r_2 - x_2'v/c} - \frac{x_1' - r_1 v/c}{r_1 - x_1'v/c}\right). \tag{4-3.21}$$

Now, by Eq. (4-3.1), $x_2' - r_2 v/c = L_2$, and by Eq. (4-3.2), $x_1' - r_1 v/c = L_1$. Substituting L_2 and L_1 into Eq. (4-3.21) and transforming the denominators to the present position quantities by means of Eqs. (4-3.7)-(4-3.12), just as we did in Eq. (4-3.6), we finally obtain

$$E_y = \frac{\lambda}{4\pi\varepsilon_0 R^2}\left[\frac{L_2}{(L_2^2/R^2 + 1 - v^2/c^2)^{1/2}} - \frac{L_1}{(L_1^2/R^2 + 1 - v^2/c^2)^{1/2}}\right]. \tag{4-3.22}$$

The magnetic field. Although we could find the magnetic field of the moving line charge from Eq. (2-2.2) or from Eq. (2-2.5), it is much simpler to find it from the electric field of the charge. According to Eq. (3-2.5), the magnetic field **H** of any uniformly moving charge distribution is always

$$\mathbf{H} = \varepsilon_0 \mathbf{v} \times \mathbf{E}, \tag{4-3.23}$$

where **E** is the electric field of the moving charge distribution. Since $\mathbf{v} = -v\mathbf{i}$, the only non-vanishing component of the cross product in Eq. (4-3.23) is the z component involving E_y only. Substituting **v** and Eq. (4-3.22) into Eq. (4-3.23) and denoting λv as the current I, we obtain

$$\mathbf{H} = \mathbf{k}\frac{I}{4\pi R^2}\left[\frac{L_1}{(L_1^2/R^2+1-v^2/c^2)^{1/2}} - \frac{L_2}{(L_2^2/R^2+1-v^2/c^2)^{1/2}}\right]. \tag{4-3.24}$$

4-4. The Electric Field of a Point Charge in Arbitrary Motion

As before, we consider a constant charge distribution of total charge q and density ρ confined to a small rectangular prism (Fig. 4.6) whose center is located at the point x', y' in the xy plane of a rectangular system of coordinates, and whose sides l, a, and b are parallel to the x, y, and z axis, respectively. The point of observation is at the origin. The distance of the center of the prism from the point of observation (the origin) is $r_0 \gg a, b, l$, so that the prism constitutes a point charge.[2] We shall assume that at the retarded time t' the center of the prism moves with velocity **v** in the negative x direction and has an acceleration $\dot{\mathbf{v}}$.

For a given present time t, the retarded times associated with different points of the prism are different, corresponding to different retarded distances of these points from the point of observation. Therefore the retarded velocities of the different

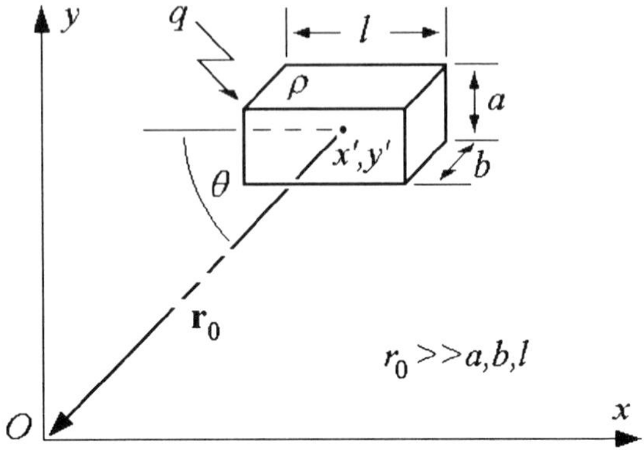

Fig. 4.6 A charge of uniform density ρ is confined to a small rectangular prism. The charge constitutes a point charge when viewed from a distance large compared to its linear dimensions.

points of the prism are also different. If the prism is sufficiently far from the point of observation, which we assume to be the case, the difference between the retarded times corresponding to different points of the prism is very small, and therefore the retarded acceleration of the prism may be assumed to have the same value $\dot{\mathbf{v}}$ for all points of the prism, even if in reality the acceleration is variable. Therefore the velocities of the different points of the prism can be calculated from velocity formulas for motion with constant acceleration.

As we shall presently see, in addition to the velocity of the center of the prism, we only need the velocities of the right, left, top, and bottom surfaces of the prism. Let the distances of these surfaces from the point of observation be r_1, r_2, r_3, and r_4, as shown in Fig. 4.7. The time interval between the retarded time for the center of the prism and for its left or right surface is then approximately $(r_1-r_2)/2c$ (see Section 3.1), and the time interval between the retarded time for the center of the prism and for its top or bottom surface is approximately $(r_3-r_4)/2c$. Therefore the

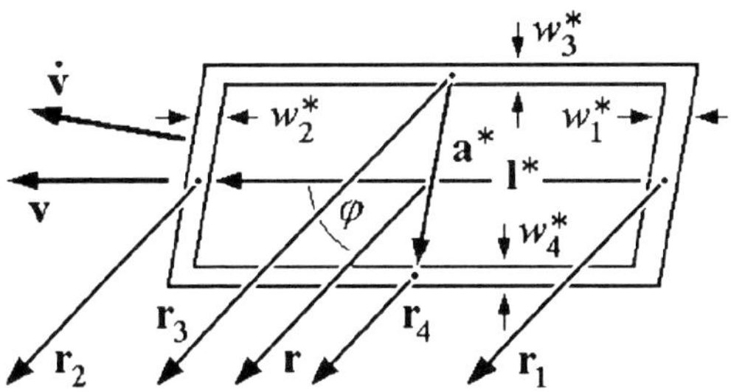

Fig. 4.7 When the charge shown in Fig. 4.6 is in a state of accelerated motion and is at a retarded position, its apparent length, shape, and thickness of its surface layers are no longer the same as for the stationary charge. The distances from the center of the charge and from four surface layers to the point of observation are represented by the vectors $\mathbf{r}$, $\mathbf{r}_1$, $\mathbf{r}_2$, $\mathbf{r}_3$, *and* $\mathbf{r}_4$. *All five* $\mathbf{r}$'s *meet at the point of observation (origin of coordinates). The acceleration vector is in the xy plane.*

(approximate) retarded velocities of the right, left, top, and bottom surfaces of the prism are, respectively, $\mathbf{v}_1 = \mathbf{v} - \dot{\mathbf{v}}(r_1 - r_2)/2c$, $\mathbf{v}_2 = \mathbf{v} + \dot{\mathbf{v}}(r_1 - r_2)/2c$, $\mathbf{v}_3 = \mathbf{v} - \dot{\mathbf{v}}(r_3 - r_4)/2c$, and $\mathbf{v}_4 = \mathbf{v} + \dot{\mathbf{v}}(r_3 - r_4)/2c$.

As was explained in Section 3-1, the apparent size and shape of the prism in its retarded position is not the same as that of the prism when it is at rest. In particular, if the prism moves in the $-x$ direction, the prism appears to be longer, it appears to be slanted, and the effective volume of the prism and of its surface layers changes (Fig. 4.7). As a result, the following geometrical relations hold for the moving prism at its retarded position:

The apparent length of the prism is, by Eq. (3-1.7),

$$l^* = \frac{l}{1 - \mathbf{r} \cdot \mathbf{v}/rc}. \tag{4-4.1}$$

The apparent volume of the prism is, by Eq. (3-1.8),

$$(abl)^* = \frac{abl}{1 - \mathbf{r} \cdot \mathbf{v}/rc} . \tag{4-4.2}$$

By the same equations, the apparent volume of the right surface layer (distance r_1 from the origin) is

$$(abw)_1^* = \frac{abw}{1 - \mathbf{r}_1 \cdot \mathbf{v}_1/r_1 c} ; \tag{4-4.3}$$

the apparent volume of the left surface layer (distance r_2 from the origin) is

$$(abw)_2^* = \frac{abw}{1 - \mathbf{r}_2 \cdot \mathbf{v}_2/r_2 c} ; \tag{4-4.4}$$

the apparent volume of the top surface layer (distance r_3 from the origin) is

$$(lbw)_3^* = \frac{lbw}{1 - \mathbf{r}_3 \cdot \mathbf{v}_3/r_3 c} ; \tag{4-4.5}$$

and the apparent volume of the bottom surface layer (distance r_4 from the origin) is

$$(lbw)_4^* = \frac{lbw}{1 - \mathbf{r}_4 \cdot \mathbf{v}_4/r_4 c} . \tag{4-4.6}$$

We shall find the electric field of our accelerating point charge by using Eq. (2-2.1)

$$\mathbf{E} = -\frac{1}{4\pi\varepsilon_0} \int \frac{\left[\nabla'\rho + \dfrac{1}{c^2}\dfrac{\partial \mathbf{J}}{\partial t}\right]}{r} dV' . \tag{2-2.1}$$

Consider first the contribution of the gradient of the charge density, $\nabla'\rho$, to the field. Since ρ is constant within the charge, $\nabla'\rho = 0$ within it, so that the only contribution to $\nabla'\rho$ comes from the surface layer of the charge, where ρ changes from 0 (outside the charge) to ρ (inside the charge). Let the actual thickness of the surface layer of the charge be w. Taking into account that $\nabla'\rho$ represents the rate of change of ρ in the direction of the greatest rate of change, we then have $\nabla'\rho = (\rho/w)\mathbf{n}_{in}$, where $\mathbf{n}_{in}$ is a unit vector normal to the surface layer and pointing *into* the charge.[7] Since the center of the charge is in the xy plane ($z' = 0$), the integrals over the two surface layers parallel to the xy plane cancel each other, because $\nabla'\rho$ for one of the layers is opposite to that for the other layer, while r is the same for both layers. Thus, as far as $\nabla'\rho$ is concerned, only the four integrals over the layers parallel to the xz and yz planes remain. Referring to Figs. 4.6 and 4.7, they are the right, left, top, and bottom surface layers, and $\nabla'\rho$ associated with these surface layers is, respectively $-(\rho/w)\mathbf{i}$, $(\rho/w)\mathbf{i}$, $-(\rho/w)\mathbf{j}$, and $(\rho/w)\mathbf{j}$ (these are the same relations that we used for finding the electric field of a uniformly moving point charge in Section 4.1).

Assuming that r_1, r_2, r_3, and r_4 are much larger than l^*, we can replace the integrals over the four layers by the products of the integrands and the retarded volumes of the layers, which gives

$$\mathbf{E} = -\frac{1}{4\pi\varepsilon_0}\left[\frac{\rho/w}{r_1}(abw)_1^*(-\mathbf{i}) + \frac{\rho/w}{r_2}(abw)_2^*\mathbf{i} + \frac{\rho/w}{r_3}(lbw)_3^*w(-\mathbf{j}) + \frac{\rho/w}{r_4}(lbw)_4^*w\mathbf{j}\right] - \frac{1}{4\pi\varepsilon_0 c^2}\int\frac{[\partial\mathbf{J}/\partial t]}{r}dv'. \quad (4\text{-}4.7)$$

Let us designate the part of Eq. (4-4.7) which explicitly depends on ρ as $\mathbf{E}_\rho$. Using Eqs. (4-4.3)-(4-4.6) and cancelling w, we can write then

$$\mathbf{E}_\rho = -\frac{\rho}{4\pi\varepsilon_0}\left[\left(\frac{1}{r_2\{1-\mathbf{r}_2\cdot\mathbf{v}_2/r_2c\}} - \frac{1}{r_1\{1-\mathbf{r}_1\cdot\mathbf{v}_1/r_1c\}}\right)ab\mathbf{i} + \left(\frac{1}{r_4\{1-\mathbf{r}_4\cdot\mathbf{v}_4/r_4c\}} - \frac{1}{r_3\{1-\mathbf{r}_3\cdot\mathbf{v}_3/r_3c\}}\right)bl\mathbf{j}\right]. \quad (4\text{-}4.8)$$

The differences of the fractions in this equation are simply the increments of the function $1/(r - \mathbf{r}\cdot\mathbf{v}/c)$ associated with the displacement of the tail of $\mathbf{r}$ over a small distance represented by the vector **l*** [in the **i** component of Eq. (4-4.8)] and by the vector **a*** [in the **j** component of Eq. (4-4.8)]. Therefore, just as we did in the case of Eq. (4-1.7), we can write Eq. (4-4.8) as

$$\mathbf{E}_\rho = -\frac{\rho b}{4\pi\varepsilon_0}\left\{\left[\left(\nabla'\frac{1}{r-\mathbf{r}\cdot\mathbf{v}/c}\right)\cdot\mathbf{l}^*\right]a\mathbf{i}+\left[\left(\nabla'\frac{1}{r-\mathbf{r}\cdot\mathbf{v}/c}\right)\cdot\mathbf{a}^*\right]l\mathbf{j}\right\}. \quad (4\text{-}4.9)$$

Using Eqs. (3-1.16), (3-1.14), and (3-1.15), we now have

$$\begin{aligned}\mathbf{E}_\rho = \frac{\rho b}{4\pi\varepsilon_0}&\left[\left(\frac{\mathbf{r}-r\mathbf{v}/c+(\mathbf{r}\cdot\dot{\mathbf{v}})\mathbf{r}/c^2}{r^3(1-\mathbf{r}\cdot\mathbf{v}/rc)^2}\cdot\mathbf{i}\right)\frac{la}{1-\mathbf{r}\cdot\mathbf{v}/rc}\mathbf{i}\right.\\ &+\left(\frac{\mathbf{r}-r\mathbf{v}/c+(\mathbf{r}\cdot\dot{\mathbf{v}})\mathbf{r}/c^2}{r^3(1-\mathbf{r}\cdot\mathbf{v}/rc)^2}\cdot\mathbf{i}\right)\frac{y'v/c}{r(1-\mathbf{r}\cdot\mathbf{v}/rc)}al\mathbf{j}\\ &\left.+\left(\frac{\mathbf{r}-r\mathbf{v}/c+(\mathbf{r}\cdot\dot{\mathbf{v}})\mathbf{r}/c^2}{r^3(1-\mathbf{r}\cdot\mathbf{v}/rc)^2}\cdot\mathbf{j}\right)\frac{r-x'v/c}{r(1-\mathbf{r}\cdot\mathbf{v}/rc)}al\mathbf{j}\right].\end{aligned} \quad (4\text{-}4.10)$$

Simplifying and taking into account that $\mathbf{r}\cdot\mathbf{i} = -x'$, $\mathbf{r}\cdot\mathbf{j} = -y'$, $\mathbf{v}\cdot\mathbf{i} = -v$, $\mathbf{v}\cdot\mathbf{j} = 0$, and $\mathbf{r}\cdot\mathbf{v} = x'v$, we obtain

$$\begin{aligned}\mathbf{E}_\rho &= \frac{\rho abl}{4\pi\varepsilon_0 r^3[1-\mathbf{r}\cdot\mathbf{v}/rc]^3}\Big\{[-x'+rv/c-(\mathbf{r}\cdot\dot{\mathbf{v}})x'/c^2]\,\mathbf{i}\\ &+[-x'+rv/c-(\mathbf{r}\cdot\dot{\mathbf{v}})x'/c^2]\frac{y'v/c}{r}\mathbf{j}\\ &+[-y'-(\mathbf{r}\cdot\dot{\mathbf{v}})y'/c^2]\frac{r-x'v/c}{r}\mathbf{j}\Big\}\\ &= \frac{\rho abl}{4\pi\varepsilon_0 r^3[1-\mathbf{r}\cdot\mathbf{v}/rc]^3}[-x'\mathbf{i}-r\mathbf{v}/c-(\mathbf{r}\cdot\dot{\mathbf{v}})x'/c^2\mathbf{i}\\ &+(v^2y'/c^2)\mathbf{j}-y'\mathbf{j}-(\mathbf{r}\cdot\dot{\mathbf{v}})y'/c^2\mathbf{j}].\end{aligned} \quad (4\text{-}4.11)$$

Since we are not interested in the acceleration-independent field $\mathbf{E}_V$ (this field was found in Section 4-1), we shall drop in Eq. (4-4.11) the terms that do not contain the acceleration $\dot{\mathbf{v}}$, and shall designate the rest of the equations as $\mathbf{E}_{A\rho}$, with the subscript "A" standing for "acceleration." Noting that $\mathbf{r} = -x'\mathbf{i} - y'\mathbf{j}$, and that $\rho abl = q$, we then obtain

$$\mathbf{E}_{A\rho} = \frac{q(\mathbf{r}\cdot\dot{\mathbf{v}})\mathbf{r}}{4\pi\varepsilon_0 r^3 c^2(1-\mathbf{r}\cdot\mathbf{v}/rc)^3}. \tag{4-4.12}$$

Consider now the contribution of $\partial\mathbf{J}/\partial t$ to the field. By Eq. (3-1.4), we have

$$\frac{\partial\mathbf{J}}{\partial t} = \frac{\partial(\rho\,\mathbf{v})}{\partial t} = -(\mathbf{v}\cdot\nabla'\rho)\,\mathbf{v} + \rho\frac{\partial\mathbf{v}}{\partial t} = -(\mathbf{v}\cdot\nabla'\rho)\,\mathbf{v} + \rho\,\dot{\mathbf{v}}. \tag{4-4.13}$$

However, because the retarded velocity is different in different regions (points) of the charge, we must evaluate Eq. (4-4.13) separately for each region under consideration. There are five such regions: the interior of the charge, the right surface, the left surface, the top surface, and the bottom surface.

In the interior of the charge, $\nabla'\rho = 0$. Therefore for the interior we have

$$\frac{\partial\mathbf{J}}{\partial t} = \rho\,\dot{\mathbf{v}}. \tag{4-4.14}$$

At the right surface, $\nabla'\rho = \partial\rho/\partial x'\mathbf{i} = -\rho/w\mathbf{i}$, and the velocity is $\mathbf{v}_1$. By Eq. (4-4.13), for the right surface we therefore have

$$\frac{\partial\mathbf{J}_1}{\partial t} = -(\mathbf{v}_1\cdot\nabla'\rho)\,\mathbf{v}_1 + \rho\,\dot{\mathbf{v}}_1 = -(v_{1x}\frac{\partial\rho}{\partial x'})\mathbf{v}_1 + \rho\,\dot{\mathbf{v}}_1 = (\rho/w)\,v_{1x}\mathbf{v}_1 + \rho\dot{\mathbf{v}}_1. \tag{4-4.15}$$

or

$$\frac{\partial\mathbf{J}_1}{\partial t} = (\rho/w)(v_{1x}\mathbf{v}_1 + \rho w\dot{\mathbf{v}}_1), \tag{4-4.16}$$

and since we can make w as small as we please,

$$\frac{\partial\mathbf{J}_1}{\partial t} = (\rho/w)\,v_{1x}\mathbf{v}_1. \tag{4-4.17}$$

At the left surface, $\nabla'\rho = \partial\rho/\partial x'\mathbf{i} = \rho/w\mathbf{i}$, and the velocity is $\mathbf{v}_2$. Therefore, by the same reasoning as in the case of Eq. (4-4.16),

$$\frac{\partial\mathbf{J}_2}{\partial t} = -(\rho/w)\,v_{2x}\mathbf{v}_2. \tag{4-4.18}$$

At the top surface, $\nabla'\rho = \partial\rho/\partial y'\mathbf{j} = -\rho/w\mathbf{j}$, and the velocity is $\mathbf{v}_3$. Therefore,

$$\frac{\partial \mathbf{J}_3}{\partial t} = (\rho/w)\,v_{3y}\mathbf{v}_3. \tag{4-4.19}$$

At the bottom surface, $\nabla'\rho = \partial\rho/\partial y'\mathbf{j}=\rho/w\mathbf{j}$, and the velocity is $\mathbf{v}_4$. Therefore

$$\frac{\partial \mathbf{J}_4}{\partial t} = -(\rho/w)\,v_{4y}\mathbf{v}_4. \tag{4-4.20}$$

Let us now designate the integral in Eq. (4-4.7) as $\mathbf{E}_\mathrm{J}$. Since, by supposition, all r's for the charge (prism) are much larger than the linear dimensions of the charge, we can replace the integration by the product of the respective integrands and the volumes of the five regions that contribute to $\partial\mathbf{J}/\partial t$. Using Eqs. (4-4.14), (4-4.17)-(4-4.20) and (4-4.2)-(4-4.6), we then have

$$\begin{aligned}4\pi\varepsilon_0 c^2\mathbf{E}_\mathrm{J} &= \frac{\rho\dot{\mathbf{v}}}{r}\left(\frac{abl}{1-\mathbf{r}\cdot\mathbf{v}/rc}\right)\\ &+\frac{\rho}{r_1 w}\left(v_{1x}\mathbf{v}_1\frac{abw}{1-\mathbf{r}_1\cdot\mathbf{v}_1/r_1c}\right)-\frac{\rho}{r_2 w}\left(v_{2x}\mathbf{v}_2\frac{abw}{1-\mathbf{r}_2\cdot\mathbf{v}_2/r_2c}\right)\\ &+\frac{\rho}{r_3 w}\left(v_{3y}\mathbf{v}_3\frac{lbw}{1-\mathbf{r}_3\cdot\mathbf{v}_3/r_3c}\right)-\frac{\rho}{r_4 w}\left(v_{4y}\mathbf{v}_4\frac{lbw}{1-\mathbf{r}_4\cdot\mathbf{v}_4/r_4c}\right),\end{aligned} \tag{4-4.21}$$

or

$$\begin{aligned}4\pi\varepsilon_0 c^2\mathbf{E}_\mathrm{J} &= \frac{q\dot{\mathbf{v}}}{r(1-\mathbf{r}\cdot\mathbf{v}/rc)}+\rho ab\left(\frac{v_{1x}\mathbf{v}_1}{r_1-\mathbf{r}_1\cdot\mathbf{v}_1/c}-\frac{v_{2x}\mathbf{v}_2}{r_2-\mathbf{r}_2\cdot\mathbf{v}_2/c}\right)\\ &+\rho bl\left(\frac{v_{3y}\mathbf{v}_3}{r_3-\mathbf{r}_3\cdot\mathbf{v}_3/c}-\frac{v_{4y}\mathbf{v}_4}{r_4-\mathbf{r}_4\cdot\mathbf{v}_4/c}\right).\end{aligned} \tag{4-4.22}$$

Since the linear dimensions of the charge are very small, the difference of the fractions in the last two terms of Eq. (4-4.22) can be regarded as the total differential (increment) $df = (\partial f/\partial x')dx' + (\partial f/\partial y')dy'$ of the functions

$$\frac{v_x \mathbf{v}}{r - \mathbf{r} \cdot \mathbf{v}/c} \tag{4-4.23}$$

and

$$\frac{v_y \mathbf{v}}{r - \mathbf{r} \cdot \mathbf{v}/c} \tag{4-4.24}$$

corresponding to the displacements of the tail of $\mathbf{r}$ by $\mathbf{l}^*$ and by $\mathbf{a}^*$, respectively (see Fig. 4.7).

Using Eq. (3-1.16), noting that $\mathbf{r} = -x'\mathbf{i} - y'\mathbf{j}$, noting that $v_y = 0$ (because $\mathbf{v}$ is parallel to the x axis), and remembering that $\mathbf{v}$ and v are functions of the retarded time $t' = t - r/c$, so that $\partial\mathbf{v}/\partial x' = (\partial\mathbf{v}/\partial t')\partial t'/\partial x' = (\partial\mathbf{v}/\partial t')x'/rc = \dot{\mathbf{v}}x'/rc$ with similar expressions for $\partial\mathbf{v}/\partial y'$, $\partial v/\partial x'$, and $\partial v/\partial y'$, we have for the needed partial derivatives of the two functions

$$\begin{aligned} \frac{\partial}{\partial x'}\left(\frac{v_x \mathbf{v}}{r[1-(\mathbf{r}\cdot\mathbf{v})/rc]}\right) = {} & v_x \mathbf{v}\,\frac{-x' - rv_x/c - (\mathbf{r}\cdot\dot{\mathbf{v}})x'/c^2}{r^3[1-(\mathbf{r}\cdot\mathbf{v})/rc]^2} \\ & - \frac{(\dot{v}_x \mathbf{v} + v_x \dot{\mathbf{v}})x'}{r^2 c[1-(\mathbf{r}\cdot\mathbf{v})/rc]}, \end{aligned} \tag{4-4.25}$$

$$\frac{\partial}{\partial x'}\left(\frac{v_y \mathbf{v}}{r[1-(\mathbf{r}\cdot\mathbf{v})/rc]}\right) = -\frac{\dot{v}_y \mathbf{v} x'}{r^2 c[1-(\mathbf{r}\cdot\mathbf{v})/rc]}, \tag{4-4.26}$$

and

$$\frac{\partial}{\partial y'}\left(\frac{v_y \mathbf{v}}{r[1-(\mathbf{r}\cdot\mathbf{v})/rc]}\right) = -\frac{\dot{v}_y \mathbf{v} y'}{r^2 c[1-(\mathbf{r}\cdot\mathbf{v})/rc]}. \tag{4-4.27}$$

In evaluating Eq. (4-4.22) with the help of Eqs. (4-4.25)-(4-4.27), we shall omit from Eq. (4-4.25) terms not containing $\dot{\mathbf{v}}$, since they only contribute to the acceleration-independent field $\mathbf{E}_V$, which we already found in Section 4-1. Combining Eqs. (4-4.22), (4-4.25)-(4-4.27), (3-1.14), and (3-1.15), we then have, denoting the acceleration-dependent field as $\mathbf{E}_{JA}$,

$$
\begin{aligned}
4\pi\varepsilon_0 c^2 \mathbf{E}_{JA} &= \frac{q\dot{\mathbf{v}}}{r(1-\mathbf{r}\cdot\mathbf{v}/rc)} \\
&+\rho ab\left[\frac{-v_x\mathbf{v}(\mathbf{r}\cdot\dot{\mathbf{v}})x'}{r^3c^2(1-\mathbf{r}\cdot\mathbf{v}/rc)^2}-\frac{(\dot{v}_x\mathbf{v}+v_x\dot{\mathbf{v}})x'}{r^2c(1-\mathbf{r}\cdot\mathbf{v}/rc)}\right]\cdot\frac{l}{(1-\mathbf{r}\cdot\mathbf{v}/rc)} \\
&-\rho bl\left[\frac{\dot{v}_y\mathbf{v}\,x'}{r^2c(1-\mathbf{r}\cdot\mathbf{v}/rc)}\cdot\frac{ay'v/c}{r(1-\mathbf{r}\cdot\mathbf{v}/rc)}\right. \\
&\left.+\frac{\dot{v}_y\mathbf{v}\,y'}{r^2c(1-\mathbf{r}\cdot\mathbf{v}/rc)}\cdot\frac{a(r-\mathbf{r}\cdot\mathbf{v}/c)}{r(1-\mathbf{r}\cdot\mathbf{v}/rc)}\right],
\end{aligned}
\tag{4-4.28}
$$

or

$$
\begin{aligned}
4\pi\varepsilon_0 c^2 \mathbf{E}_{JA} &= \frac{q\dot{\mathbf{v}}}{r(1-\mathbf{r}\cdot\mathbf{v}/rc)} \\
&+\frac{q}{r^2c(1-\mathbf{r}\cdot\mathbf{v}/rc)^2}\left[\frac{-v_x\mathbf{v}(\mathbf{r}\cdot\dot{\mathbf{v}})x'}{rc(1-\mathbf{r}\cdot\mathbf{v}/rc)}-\dot{v}_x\mathbf{v}x'\right. \\
&\left.-v_x\dot{\mathbf{v}}x'-\frac{\dot{v}_y\mathbf{v}\,x'y'v}{rc}-\dot{v}_y\mathbf{v}\,y'+\frac{\dot{v}_y\mathbf{v}y'(\mathbf{r}\cdot\mathbf{v})}{rc}\right].
\end{aligned}
\tag{4-4.29}
$$

Since $\mathbf{r}\cdot\mathbf{v} = x'v = -x'v_x$ and since $-\dot{v}_x x' - \dot{v}_y y' = \dot{\mathbf{v}}\cdot\mathbf{r}$ (see Figs. 4.6 and 4.7), Eq. (4-4.29) reduces to

$$
\begin{aligned}
4\pi\varepsilon_0 c^2 \mathbf{E}_{JA} &= \frac{q\dot{\mathbf{v}}}{r(1-\mathbf{r}\cdot\mathbf{v}/rc)} \\
&+\frac{q}{r^2c(1-\mathbf{r}\cdot\mathbf{v}/rc)^2}\left[\frac{\mathbf{v}(\mathbf{r}\cdot\mathbf{v})(\mathbf{r}\cdot\dot{\mathbf{v}})}{rc(1-\mathbf{r}\cdot\mathbf{v}/rc)}+(\mathbf{r}\cdot\dot{\mathbf{v}})\mathbf{v}+(\mathbf{r}\cdot\mathbf{v})\dot{\mathbf{v}}\right],
\end{aligned}
\tag{4-4.30}
$$

which after elementary simplifications becomes

$$
\mathbf{E}_{JA} = \frac{q\dot{\mathbf{v}}}{4\pi\varepsilon_0 c^2 r(1-\mathbf{r}\cdot\mathbf{v}/rc)^2}+\frac{q(\mathbf{r}\cdot\dot{\mathbf{v}})\mathbf{v}}{4\pi\varepsilon_0 c^3 r^2(1-\mathbf{r}\cdot\mathbf{v}/rc)^3}. \tag{4-4.31}
$$

Finally, in accordance with Eq. (4-4.7), subtracting Eq. (4-4.31) from Eq. (4-4.12), we obtain for $\mathbf{E}_A$

$$\mathbf{E}_A = \frac{q(\mathbf{r}\cdot\dot{\mathbf{v}})\mathbf{r}}{4\pi\varepsilon_0 c^2 r^3(1-\mathbf{r}\cdot\mathbf{v}/rc)^3} - \frac{q(\mathbf{r}\cdot\dot{\mathbf{v}})\mathbf{v}}{4\pi\varepsilon_0 c^3 r^2(1-\mathbf{r}\cdot\mathbf{v}/rc)^3} - \frac{q\dot{\mathbf{v}}}{4\pi\varepsilon_0 c^2 r(1-\mathbf{r}\cdot\mathbf{v}/rc)^2}, \quad (4\text{-}4.32)$$

which can be written in a simpler form as

$$\mathbf{E}_A = \frac{q}{4\pi\varepsilon_0 r^3 c^2(1-\mathbf{r}\cdot\mathbf{v}/rc)^3}\left\{\mathbf{r}\times\left[\left(\mathbf{r}-\frac{r\mathbf{v}}{c}\right)\times\dot{\mathbf{v}}\right]\right\}. \quad (4\text{-}4.33)$$

The total electric field is the sum of the acceleration-independent field $\mathbf{E}_V$ given by Eq. (4-1.11) and of $\mathbf{E}_A$ given by Eq. (4-4.33). Adding Eqs. (4-1.11) and (4-4.33), we obtain for the total electric field of a point charge in arbitrary motion

$$\mathbf{E} = \frac{q}{4\pi\varepsilon_0 r^3(1-\mathbf{r}\cdot\mathbf{v}/rc)^3}\left\{\left(\mathbf{r}-\frac{r\mathbf{v}}{c}\right)\left(1-\frac{v^2}{c^2}\right)+\mathbf{r}\times\left[\left(\mathbf{r}-\frac{r\mathbf{v}}{c}\right)\times\frac{\dot{\mathbf{v}}}{c^2}\right]\right\}. \quad (4\text{-}4.34)$$

Note that $\mathbf{r}$, r, $\mathbf{v}$, v, and $\dot{\mathbf{v}}$ in this equation are retarded.

▼

Example 4-4.1 A point charge moves with constant speed along a circle of radius r (Fig 4.8). Find the electric and magnetic fields produced by the charge at the center of the circle and discuss the significance of the resulting equations for electrodynamics of atomic systems.

For circular motion $\dot{\mathbf{v}} = (v^2/r^2)\mathbf{r}$. Substituting $\dot{\mathbf{v}}$ into Eq. (4-4.34), taking into account that $\mathbf{r}\cdot\mathbf{v} = 0$, and simplifying, we obtain

$$\mathbf{E} = \frac{q}{4\pi\varepsilon_0 r^3}\left\{\mathbf{r}\left(1-\frac{v^2}{c^2}\right)-\mathbf{v}\frac{r}{c}\right\}. \quad (4\text{-}4.35)$$

Equation (4-4.35) expresses the electric field in terms of the retarded position of the charge. Let us convert this expression to the

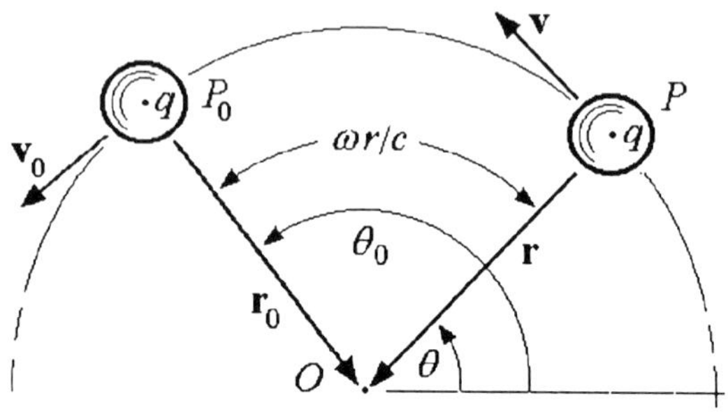

Fig. 4.8 Geometrical relations between the "present position vector" $\mathbf{r}_0$ *and the "retarded position vector"* $\mathbf{r}$ *for a point charge q moving with velocity* $\mathbf{v}$ *in a circular orbit. The field signal originates at the "retarded" point P and propagates with velocity c toward the center of the orbit O. By the time the signal reaches the center of the orbit, the charge has moved an angular distance* $\omega r/c$ *along the orbit and is at the "present position" point* P_0*. (Note: The length of the arc between P nd* P_0 *is exaggerated. Since* $v < c$*, the arc should be shorter than the radius of the orbit.)*

present position of the charge. We can do so by resolving the retarded position vector $\mathbf{r}$ and the retarded velocity vector $\mathbf{v}$ into their components along the present position vector $\mathbf{r}_0$ and the present velocity vector $\mathbf{v}_0$. Since the angle between the present position vector and the retarded position vector is $\theta_0 - \theta = \omega r/c = v/c$, where ω is the angular velocity of the charge, we obtain for the two components of $\mathbf{E}$

$$E_{r_0} = \frac{q}{4\pi\varepsilon_0 r^3}\left\{\left(1 - \frac{v^2}{c^2}\right) r\cos(v/c) + \frac{rv}{c}\sin(v/c)\right\}, \tag{4-4.36}$$

$$E_{v_0} = \frac{q}{4\pi\varepsilon_0 r^3}\left\{\left(1 - \frac{v^2}{c^2}\right) r\sin(v/c) - \frac{rv}{c}\cos(v/c)\right\}, \tag{4-4.37}$$

and for the total field

$$\mathbf{E} = \frac{q}{4\pi\varepsilon_0 r^3}\left\{\left[\left(1 - \frac{v^2}{c^2}\right)\cos(v/c) + \frac{v}{c}\sin(v/c)\right]\mathbf{r}_0 + \left[\left(1 - \frac{v^2}{c^2}\right)\frac{r}{v}\sin(v/c) - \frac{r}{c}\cos(v/c)\right]\mathbf{v}_0\right\}. \tag{4-4.38}$$

The most obvious practical application of Eq. (4-4.38) is for the case when we can neglect v/c to powers higher than 3. Expanding $\sin(v/c)$ and $\cos(v/c)$ in Eq. (4-4.38) into power series of v/c and dropping terms containing v/c to powers higher than 3, we have

$$\mathbf{E} = \frac{q}{4\pi\varepsilon_0 r^3}\left\{\left(1 - \frac{v^2}{2c^2}\right)\mathbf{r}_0 - \frac{2rv^2}{3c^3}\mathbf{v}_0\right\}. \tag{4-4.39}$$

To find the magnetic field, we apply Eq. (3-2.12) to the electric field given by Eq. (4-4.35). This gives

$$\mathbf{H} = \frac{q}{4\pi\varepsilon_0\mu_0 r^4 c}\mathbf{r}\times\left\{\mathbf{r}\left(1 - \frac{v^2}{c^2}\right) - \mathbf{v}\frac{r}{c}\right\}, \tag{4-4.40}$$

or, since $\mathbf{r} \times \mathbf{r} = 0$, and $1/\varepsilon_0\mu_0 = c^2$,

$$\mathbf{H} = \frac{q}{4\pi r^3}[\mathbf{v}\times\mathbf{r}]. \tag{4-4.41}$$

Although $\mathbf{v}$ and $\mathbf{r}$ in Eq. (4-4.41) are retarded, their cross product is not affected by conversion to the present velocity vector and present position vector of the charge, because the cross product is the same for all points of the orbit. Therefore the magnetic field given by Eq. (4-4.41) is exactly as expected from the Biot-Savart law. But Eqs. (4-4.38) and (4-4.39) for the electric field are quite unexpected. Intuitively, one would expect the field to be the Coulomb field [possibly with the factor $(1 - v^2/c^2)$] directed to the center of the orbit. Contrary to expectations, the true electric field of a point charge moving with constant speed in a circular orbit is very different from the Coulomb field: First, the field has a component parallel to the instantaneous velocity vector, and thus is *not* directed to the center of the orbit. Second, the field is not proportional to $1/r^2$. Third, the factor in the radial component of the field is $(1 - v^2/2c^2)$ rather than $(1 - v^2/c^2)$.

As far as atomic systems are concerned, it is clear from the derivations presented above that the Coulomb law cannot be used as a rigorous basis for any atomic model. The problem is that, even if the electric field of the nucleus is exactly a Coulomb field, so that the electric force exerted by the nucleus on electrons is the ordinary $1/r^2$ force, the electric force exerted by electrons on the nucleus is, by Eqs. (4-4.38) and (4-4.39), neither radial nor proportional to $1/r^2$ [the fact that Eqs. (4-4.38) and (4-4.39) have been obtained for a circular, rather than for an elliptical, orbit cannot possibly change the essence of the information provided by Eqs. (4-4.38) and (4-4.39)]. Therefore any atomic model based on Coulomb field or Coulomb potential can at best be only approximately correct, although the corrections associated with the acceleration of the electrons are clearly very small.[8]

▲

4-5. The Magnetic Field of a Point Charge in Arbitrary Motion

Although by using Eq. (2-2.2) or Eq. (2-2.5) we can find the magnetic field produced by a point charge in arbitrary motion in the same manner as we found the electric field in Section 4-4 (see Example 4-5.1), it is much easier to find it from the known electric field by using Eq. (3-2.12).

Applying Eq. (3-2.12) to Eq. (4-4.33) and using $\varepsilon_0\mu_0 = 1/c^2$, we obtain for the acceleration part of the magnetic field after elementary simplifications

$$\mathbf{H}_A = \frac{q}{4\pi r^2 c(1 - \mathbf{r}\cdot\mathbf{v}/rc)^2}\left[\frac{(\mathbf{v}\times\mathbf{r})(\mathbf{r}\cdot\dot{\mathbf{v}})}{rc(1 - \mathbf{r}\cdot\mathbf{v}/rc)} + \dot{\mathbf{v}}\times\mathbf{r}\right]. \quad (4\text{-}5.1)$$

Applying Eq. (3-2.12) to Eq. (4-4.34) and using $\varepsilon_0\mu_0 = 1/c^2$, we obtain for the total magnetic field after elementary simplifications

$$\mathbf{H} = \frac{qc}{4\pi r^2(1-\mathbf{r}\cdot\mathbf{v}/rc)^2}\left[\frac{1-v^2/c^2+\mathbf{r}\cdot\dot{\mathbf{v}}/c^2}{r(1-\mathbf{r}\cdot\mathbf{v}/rc)}(\mathbf{v}\times\mathbf{r})+\frac{\dot{\mathbf{v}}\times\mathbf{r}}{c}\right]. \quad (4\text{-}5.2)$$

▼

Example 4-5.1 Find the magnetic field of an accelerating point charge shown in Figs. 4.6 and 4.7 by using Eq. (2-2.2).

Since $\mathbf{J} = \rho\mathbf{v}$, $\nabla' \times \mathbf{J} = \nabla' \times \rho\mathbf{v} = \nabla'\rho \times \mathbf{v} + \rho\nabla' \times \mathbf{v}$. But $\mathbf{v}$ is not a point function (there is no "velocity field"), and therefore $\nabla' \times \mathbf{v} = 0$ and $\nabla' \times \mathbf{J} = \nabla'\rho \times \mathbf{v}$. As we already know from Sections 4-1 and 4-4, $\nabla'\rho$ for our charge is only different from zero at the surface layers of the charge. Therefore the only contribution to the integral in Eq. (2-2.2) comes from the right, left, top, and bottom surface layers, where $\nabla'\rho$ is $-(\rho/w)\mathbf{i}$, $(\rho/w)\mathbf{i}$, $-(\rho/w)\mathbf{j}$, and $(\rho/w)\mathbf{j}$, respectively (by symmetry, the contributions of the front and back surface layers cancel). Since $[\nabla' \times \mathbf{J}]$ in the integral of Eq. (2-2.2) is retarded, the velocity in the expression $[\nabla'\rho \times \mathbf{v}]$ is the retarded velocity of each surface under consideration. By supposition, the distances from the charge to the point of observation is much larger than l^*. Therefore the integral in Eq. (2-2.2) can be replaced by the integrand and the volume of integration (the respective volumes of the surface layers). Substituting into $[\nabla' \times \mathbf{J}] = [\nabla'\rho \times \mathbf{v}] = -[\mathbf{v} \times \nabla'\rho]$ the above expressions for $\nabla'\rho$, and using Eqs. (2-2.2) and (4-4.3)-(4-4.6), we then have

$$\mathbf{H} = \frac{\rho}{4\pi w}\left[\frac{abw(\mathbf{v}_1\times\mathbf{i})}{r_1\{1-\mathbf{r}_1\cdot\mathbf{v}_1/r_1c\}} - \frac{abw(\mathbf{v}_2\times\mathbf{i})}{r_2\{1-\mathbf{r}_2\cdot\mathbf{v}_2/r_2c\}} + \frac{blw(\mathbf{v}_3\times\mathbf{j})}{r_3\{1-\mathbf{r}_3\cdot\mathbf{v}_3/r_3c\}} - \frac{blw(\mathbf{v}_4\times\mathbf{j})}{r_4\{1-\mathbf{r}_4\cdot\mathbf{v}_4/r_4c\}}\right], \quad (4\text{-}5.3)$$

or

$$\mathbf{H} = \frac{\rho}{4\pi}\left[\left(\frac{\mathbf{v}_1}{[r_1 - \mathbf{r}_1 \cdot \mathbf{v}_1/c]} - \frac{\mathbf{v}_2}{[r_2 - \mathbf{r}_2 \cdot \mathbf{v}_2/c]}\right) \times \mathbf{i}ab \right.$$
$$\left. + \left(\frac{\mathbf{v}_3}{[r_3 - \mathbf{r}_3 \cdot \mathbf{v}_3/c]} - \frac{\mathbf{v}_4}{[r_4 - \mathbf{r}_4 \cdot \mathbf{v}_4/c]}\right) \times \mathbf{j}bl\right]. \quad (4\text{-}5.4)$$

The differences of the fractions in Eq. (4-5.4), just as before in Eq. (4-4.22), are the increments of the functions given by Eqs. (4-4.23) and (4-4.24), except that v_x and v_y are now absent. By Eqs. (4-4.25) and (4-4.26), taking into account that $v_y = 0$, the corresponding partial derivatives are

$$\frac{\partial}{\partial x'}\left(\frac{\mathbf{v}}{r - \mathbf{r} \cdot \mathbf{v}/c}\right) = \mathbf{v}\frac{-x' - rv_x/c - (\mathbf{r} \cdot \dot{\mathbf{v}})x'/c^2}{r^3(1 - \mathbf{r} \cdot \mathbf{v}/rc)^2} - \frac{\dot{\mathbf{v}}x'}{r^2c(1 - \mathbf{r} \cdot \mathbf{v}/rc)}, \quad (4\text{-}5.5)$$

and

$$\frac{\partial}{\partial y'}\left(\frac{\mathbf{v}}{r - \mathbf{r} \cdot \mathbf{v}/c}\right) = \mathbf{v}\frac{-y' - (\mathbf{r} \cdot \dot{\mathbf{v}})y/c^2}{r^3(1 - \mathbf{r} \cdot \mathbf{v}/rc)^2} - \frac{\dot{\mathbf{v}}y'}{r^2c(1 - \mathbf{r} \cdot \mathbf{v}/rc)}. \quad (4\text{-}5.6)$$

In evaluating Eq. (4-5.4) with the help of Eqs. (4-5.5) and (4-5.6), we shall omit from Eqs. (4-5.5) and (4-5.6) terms not containing $\dot{\mathbf{v}}$, since they only contribute to $\mathbf{H}_V$, (the magnetic field of a uniformly moving charge), which we do not need. Combining Eqs. (4-5.4), (4-5.5), (4-5.6), (3-1.14), and (3-1-15), we then have for the acceleration-dependent field

$$\mathbf{H}_A = \frac{\rho}{4\pi}\left[\left(\frac{-\mathbf{v}(\mathbf{r} \cdot \dot{\mathbf{v}})x'}{r^3c^2(1 - \mathbf{r} \cdot \mathbf{v}/rc)} - \frac{\dot{\mathbf{v}}x'}{r^2c(1 - \mathbf{r} \cdot \mathbf{v}/rc)}\right) \times \mathbf{i} \cdot \frac{abl}{1 - \mathbf{r} \cdot \mathbf{v}/rc}\right.$$
$$+ \left(\frac{-\mathbf{v}(\mathbf{r} \cdot \dot{\mathbf{v}})x'}{r^3c^2(1 - \mathbf{r} \cdot \mathbf{v}/rc)} - \frac{\dot{\mathbf{v}}x'}{r^2c(1 - \mathbf{r} \cdot \mathbf{v}/rc)}\right) \times \mathbf{j} \cdot \frac{ably'v/c}{r(1 - \mathbf{r} \cdot \mathbf{v}/rc)} \quad (4\text{-}5.7)$$
$$\left. + \left(\frac{-\mathbf{v}(\mathbf{r} \cdot \dot{\mathbf{v}})y'}{r^3c^2(1 - \mathbf{r} \cdot \mathbf{v}/rc)} - \frac{\dot{\mathbf{v}}y'}{r^2c(1 - \mathbf{r} \cdot \mathbf{v}/rc)}\right) \times \mathbf{j} \cdot \frac{ably'(1 - \mathbf{r} \cdot \mathbf{v}/rc)}{r(1 - \mathbf{r} \cdot \mathbf{v}/rc)}\right].$$

Expanding Eq. (4-5.7), taking into account that $\mathbf{v} \times \mathbf{i} = 0$, and simplifying, we obtain

$$\mathbf{H}_A = \frac{q}{4\pi r^2 c(1 - \mathbf{r} \cdot \mathbf{v}/rc)^2}\left[-(\dot{\mathbf{v}} \times \mathbf{i})x' + \frac{-\mathbf{v}(\mathbf{r} \cdot \dot{\mathbf{v}})y'}{rc(1 - \mathbf{r} \cdot \mathbf{v}/rc)} \times \mathbf{j} - \dot{\mathbf{v}} \times \mathbf{j}y'\right]. \tag{4-5.8}$$

But $\mathbf{i}x' + \mathbf{j}y' = -\mathbf{r}$, and $\mathbf{v} \times \mathbf{j}y' = -\mathbf{v} \times \mathbf{r}$ (because $\mathbf{v}$ is parallel to the x axis). Therefore Eq. (4-5.8) can be written as

$$\mathbf{H}_A = \frac{q}{4\pi r^2 c(1 - \mathbf{r} \cdot \mathbf{v}/rc)^2}\left[\dot{\mathbf{v}} \times \mathbf{r} + \frac{(\mathbf{v} \times \mathbf{r})(\mathbf{r} \cdot \dot{\mathbf{v}})}{rc(1 - \mathbf{r} \cdot \mathbf{v}/rc)}\right]. \tag{4-5.9}$$

The total magnetic field of an accelerating point charge is the sum of Eq. (4-2.1), representing the magnetic field of a uniformly moving point charge, and Eq. (4-5.9), representing the effect of the acceleration of the charge on the field. Adding Eqs. (4-2.1) and (4-5.9), we obtain

$$\mathbf{H} = \frac{q}{4\pi r^2(1 - \mathbf{r} \cdot \mathbf{v}/rc)^2}\left[\frac{1 - v^2/c^2 + (\mathbf{r} \cdot \dot{\mathbf{v}})/c^2}{r(1 - \mathbf{r} \cdot \mathbf{v}/rc)}(\mathbf{v} \times \mathbf{r}) + \frac{\dot{\mathbf{v}} \times \mathbf{r}}{c}\right]. \tag{4-5.10}$$

Observe that Eqs. (4-5.9) and (4-5.10) express the magnetic field in terms of the retarded position of the charge.

▲

4-6. Electric and Magnetic Potentials of a Moving Point Charge

Electric and magnetic potentials produced by a moving point charge q can be easily obtained from Eqs. (2-4.5) and (2-4.2).

A "point charge" is a charge distribution viewed from a distance large compared to the linear dimensions of the charge distribution. Therefore, for a point charge, the distance r in the integrals of Eqs. (2-4.5) and (2-4.2) may be considered the same for all volume elements of the charge, and therefore each integral may be replaced by the product of the integrand and the retarded volume of the charge $\Delta V'$.

By Eqs. (2-4.5) and (3-1.8), we then have for the electric scalar potential of a moving point charge

$$\varphi = \left[\frac{\rho}{4\pi\varepsilon_0 r}\Delta V'\right]_{ret} = \frac{\rho\Delta V'}{4\pi\varepsilon_0 r(1 - \mathbf{r}\cdot\mathbf{v}/rc)} \tag{4-6.1}$$

or, replacing $\rho\Delta V'$ by q,

$$\varphi = \frac{q}{4\pi\varepsilon_0 r(1 - \mathbf{r}\cdot\mathbf{v}/rc)}. \tag{4-6.2}$$

From Eqs. (2-4.2) and (3-1.8) we similarly have for the magnetic vector potential of a moving point charge

$$\mathbf{A} = \left[\frac{\mu_0\mathbf{J}}{4\pi r}\Delta V'\right]_{ret} = \frac{\mu_0\mathbf{J}\Delta V'}{4\pi r(1 - \mathbf{r}\cdot\mathbf{v}/rc)}, \tag{4-6.3}$$

and since $\mathbf{J} = \rho\mathbf{v}$,

$$\mathbf{A} = \frac{\mu_0 q\mathbf{v}}{4\pi r(1 - \mathbf{r}\cdot\mathbf{v}/rc)}. \tag{4-6.4}$$

Equations (4-6.2) and (4-6.4) are called the *Liénard-Wiechert potentials.*[9,10] They express the potentials of a moving point charge in terms of the retarded position of the charge. If the charge moves with constant velocity, Liénard-Wiechert potentials can be converted to the present position of the charge. Transforming the denominators of Eqs. (4-6.2) and (4-6.4) with the help of Eq. (3-1.27), we obtain for a point charge moving with constant velocity

$$\varphi = \frac{q}{4\pi\varepsilon_0 r_0[1 - (v^2/c^2)\sin^2\theta]^{1/2}} \tag{4-6.5}$$

and

$$\mathbf{A} = \frac{\mu_0 q\mathbf{v}}{4\pi r_0[1 - (v^2/c^2)\sin^2\theta]^{1/2}}, \tag{4-6.6}$$

where r_0 is the present position radius vector, and θ is the angle between $\mathbf{v}$ and $\mathbf{r}_0$.

▼

Example 4-6.1 Equations (4-6.5) and (4-6.6) represent the "instantaneous" potential of a uniformly moving point charge. Since the charge is moving, the potentials change as the time goes by. How should they be written to show explicitly their time dependence?

Assuming that the charge moves in the negative x direction, the x coordinate of the charge diminishes with time according to $x' - vt$, where x' is the value of the x coordinate at $t = 0$. Expressing the denominators in Eqs. (4-6.5) and (4-6.6) in terms of Cartesian coordinates by means of Eq. (3-1.26), and replacing x' by $x' - vt$, we obtain the time-dependent expressions for the potentials

$$\varphi = \frac{q}{4\pi\varepsilon_0[(x' - vt)^2 + (1 - v^2/c^2)y'^2]^{1/2}} \tag{4-6.7}$$

and

$$\mathbf{A} = \frac{\mu_0 q\mathbf{v}}{4\pi\varepsilon_0[(x' - vt)^2 + (1 - v^2/c^2)y'^2]^{1/2}}. \tag{4-6.8}$$

▲

4-7. How Accurate are the Equations for the Fields and Potentials Obtained in this Chapter?

The equations for the electric and magnetic fields of a point charge in arbitrary motion were first derived in 1898 by A. Liénard[9] from the potentials which we now call the Liénard-Wiechert potentials [Eqs. (4-6.2) and (4-6.4)]. These potentials were first derived by Liénard in 1898 and later by Wiechert in 1900.[9,10] Both Liénard and Wiechert obtained the potentials from the retarded integrals for the electric and magnetic potentials of a time-dependent charge distribution in a manner similar to our derivations presented in Section 4-6.

Liénard invented a special method for integrating retarded potential integrals for the case of a charge distribution of "very small dimensions." The essence of the method was that, because of the motion of the charge, the region of space from which the charge "sends" electric and magnetic field signals is not the same as the volume occupied by the stationary charge. According to Liénard, if the region occupied by the stationary charge is Ω, then the integration is to be extended over the region Ω/[1 - (*u*/*V*) cos (*u*,*r*)], "en prenant puor *u* et *r* une valeur moyenne," that is, by using average values for the velocity of the charge *u* and for the distance from the charge to the point of observation *r* (Liénard used *V* for the velocity of light). One should note that Liénard did not specify how these average values were supposed to be determined, and that, by using an "average value" for the velocity of the moving charge, he eliminated the need for taking into account a possible acceleration of the charge. Assuming then that the charge was "concentrated" at a "single point," Liénard obtained his "point charge" potentials.

Wiechert's derivation was essentially the same as that of Liénard. However, instead of using the average values for the velocity and distance, he simply factored out 1/*r* from under the integral sign because, according to him, "die Variation des Nenners *r* kommt bei unendlich kleinen Dimensionen nicht in Betrach," that is, because in the case of the infinitesimal volume of the charge, *r* could be regarded as constant over the volume of integration.

It is clear that since Liénard used average values of the integrand in obtaining his potentials, the potentials could not be exact. And it is also clear that Wiechert was wrong when he referred to the volume of integration as "infinitesimal." Even if the actual volume of the charge is "infinitesimal," the volume of integration is not – in fact, according to Eq. (3-1.8), it can be *infinitely large*, if the velocity of the charge is equal to the velocity of light and if the charge moves toward the point of observation!

The conventional derivations of potentials and fields of moving electric charges now used in most textbooks on electricity and magnetism are basically the same as those used by Liénard and therefore are subject to the same misgivings.

Our derivations of the electric and magnetic fields of a moving point charge presented in this chapter differ in two important aspects from the conventional derivations of these fields: (1) the fields are derived directly from the general field equations for an arbitrary time-dependent charge and current distribution, and (2) the derivations clearly reveal the physical effects responsible for the characteristic properties of the fields. In contrast, the conventional derivations, based on the Liénard-Wiechert potentials, hide these physical effects behind a physically obscure mathematical procedure required for transforming the potentials into the fields.[11] It is difficult to ascertain the range of validity of Eqs. (4-4.34) and (4-5.2) on the basis of conventional derivation. But our direct derivations show very clearly what restrictions apply to these equations and how the restrictions originate.

In obtaining the expressions for $\mathbf{E}_A$ and $\mathbf{H}_A$ we used several approximations. Our first approximation was the replacement of the integrals in Eqs. (2-2.1) and (2-2.2) by the products of the integrands and the volumes of integration. This can only be done if the relation $r \gg l^*$ is satisfied. Therefore, by Eq. (3-1.7), all our calculations are subject to the restrictions

$$r \gg \frac{l}{1-\mathbf{r}\cdot\mathbf{v}/rc} = \frac{l}{1-(v/c)\cos\phi}, \tag{4-7.1}$$

where l is the length of the "point charge," $\mathbf{v}$ is the velocity of the charge, $\mathbf{r}$ is the retarded position vector joining the charge with the point of observation, and ϕ is the angle between $\mathbf{v}$ and $\mathbf{r}$. Since Eq. (4-7.1) must hold for all values of ϕ, including $\phi = 0$, the velocity of the charge is subject to the restriction

$$v < c(1 - l/r). \tag{4-7.2}$$

Consider now the approximations that we used for taking into account the acceleration of the charge. The retarded time intervals between the center and the right-left and top-bottom surfaces of the charge are $(r_1 - r_2)/2c \approx (l \cos\phi)/[2c(1 - \mathbf{r}\cdot\mathbf{v}/rc)]$ and $(r_3 - r_4)/2c \approx (a \sin\phi)/[2c(1 - \mathbf{r}\cdot\mathbf{v}/rc)]$, respectively (see Figs. 4.7, 3.2, and 4.3).[12] For Eq. (4-7.1) to hold, the increment in the velocity of the charge during these time intervals must be less than $c - v$. Hence the restrictions on the acceleration of the charge in the direction of the x axis is

$$\dot{v}_x(r_1 - r_2)/2c < c - v, \tag{4-7.3}$$

or

$$\dot{v}_x < \frac{2(c-v)(c-v\cos\phi)}{l\cos\phi}. \tag{4-7.4}$$

A similar restriction applies to the acceleration in any other direction. Since the largest possible value for $\cos\varphi$ and $\sin\varphi$ is 1, we obtain from Eq. (4-7.4) for the general case of the acceleration $\dot{\mathbf{v}}$

$$\dot{v} < \frac{2(c-v)^2}{L}, \tag{4-7.5}$$

where L is the length of the "point charge" in the direction of the acceleration.

References and Remarks for Chapter 4

1. The calculation that follows is similar to the calculation presented in Oleg D. Jefimenko, "Direct calculation of the electric and magnetic fields of an electric point charge moving with constant velocity," Am. J. Phys. **62**, 79-85 (1994). This article also contains an example of the use of the time-independent counterpart of Eqs.

(2-2.1) and (2-2.2) for calculating the electric field of a stationary point charge and the magnetic field of a stationary current.

2. A "point charge" is by definition any charge distribution viewed from a distance large compared with the linear dimensions of this distribution, similar to the term "light point," which is frequently used in reference to stars. In neither case does the word "point" describe the structure or the constitution of the object; instead, it reflects the attitude of the observer toward this object. See Oleg D. Jefimenko, *Electricity and Magnetism*, 2nd ed., (Electret Scientific, Star City, 1989) pp. 95-98.

3. One may think that by choosing the charge in the shape of a rectangular prism we limit the generality of our derivations. This is not so. Any charge distribution can be regarded as being composed of charges confined to small rectangular prisms: this is exactly what we do when we perform integration over a volume element (rectangular prism!) $dV' = dx'\, dy'\, dz'$.

4. The increment dU of any scalar function $U\,(x, y, z,)$ associated with the displacement $d\mathbf{l} = x\mathbf{i} + y\mathbf{j} + z\mathbf{k}$ is $dU = \nabla U \cdot d\mathbf{l}$ [see, for example, Oleg D. Jefimenko, *Electricity and Magnetism*, 2nd ed., (Electret Scientific, Star City, 1989) pp. 36-38]. Since in the case under consideration the displacements $\mathbf{l}^*$ and $\mathbf{a}^*$ are very small (they represent the length and width of a "point charge"), these displacements can be treated as differentials.

5. Oliver Heaviside, "The Electromagnetic Effects of a Moving Charge," The Electrician **22**, 147-148 (1888); Oliver Heaviside, "On the Electromagnetic Effects due to the Motion of Electricity Through a Dielectric," Phil. Mag. **27**, 324-339 (1889).

6. For another (similar) derivation of the magnetic field of a uniformly moving point charge see Ref. 1.

7. Observe that Eq. (2-2.1) contains a *retarded gradient* of ρ and a *retarded time derivative* of $\mathbf{J}$, rather than a gradient of *retarded* ρ and a time derivative of *retarded* $\mathbf{J}$. This means that the gradient and the time derivative must be determined for the unretarded (stationary) ρ and $\mathbf{J}$ but must be used at the retarded position of the moving charge.

8. The same considerations apply to calculations of planetary motions based on Newton's gravitational law. See Chapter 11.

9. A. Liénard, "Champ électrique et magnetique produit par une charge électrique concentré en un point et animé d'un mouvement quelconque," L'Éclariage élect. **16**, 5-14, 53-59, 106-112 (1898).
10. E. Wiechert, "Elektrodynamische Elementargesetze," Archives Néerlanaises (2) **5**, 549-573 (1900).
11. See, for example, David J. Griffiths, *Introduction to Electrodynamics*, 2nd ed., (Prentice-Hall, Englewood Cliffs, NJ, 1989), pp. 416-426.
12. To simplify the calculations, we assume here that α in Fig. 4.3 is $\pi/2$.

5

ELECTRIC AND MAGNETIC FIELDS AND POTENTIALS OF AN ARBITRARY CHARGE DISTRIBUTION MOVING WITH CONSTANT VELOCITY

Electric and magnetic fields and potentials produced by any time-independent stationary charge and current distribution can be calculated with relative ease by a variety of methods. But calculating fields of time-dependent charge and current distributions, and the fields of moving charge distributions in particular, still remains a formidable task. In this chapter we shall obtain general formulas that allow one to determine the fields and potentials of any uniformly moving charge distribution directly and simply in terms of present time integrals that are not much different from the integrals for fields of stationary charges.

5-1. Converting Retarded Field Integrals for Uniformly Moving Charge Distributions into Present-Time (Present-Position) Integrals

As we already know from Chapters 2 and 3, electric and magnetic fields of moving charge distributions can be found from the retarded integrals

$$\mathbf{E} = -\frac{1}{4\pi\varepsilon_0}\int \frac{\left[\nabla'\rho + \frac{1}{c^2}\frac{\partial \mathbf{J}}{\partial t}\right]}{r} dV' \tag{5-1.1}$$

and

$$\mathbf{H} = \frac{1}{4\pi}\int \frac{[\nabla' \times \mathbf{J}]}{r} dV' \tag{5-1.2}$$

or from

$$\mathbf{E} = \frac{1}{4\pi\varepsilon_0}\int\left\{\frac{[\rho]}{r^3} + \frac{1}{r^2 c}\left[\frac{\partial\rho}{\partial t}\right]\right\}\mathbf{r}\,dV' - \frac{1}{4\pi\varepsilon_0 c^2}\int\frac{1}{r}\left[\frac{\partial \mathbf{J}}{\partial t}\right]dV' \tag{5-1.3}$$

and

$$\mathbf{H} = \frac{1}{4\pi}\int\left\{\frac{[\mathbf{J}]}{r^3} + \frac{1}{r^2 c}\left[\frac{\partial \mathbf{J}}{\partial t}\right]\right\}\times \mathbf{r}\,dV'. \tag{5-1.4}$$

We shall presently show that for time-independent charge distributions moving with constant velocity, these integrals can be converted to the "present" position of the charge distribution, so that the integration is performed not over the retarded, or effective, volume (see Section 3-1), but over the real volume that the charge distribution occupies at the moment t for which the fields are being determined.

The conversion is based on certain properties and relations involving retarded integrals and retarded quantities which are reviewed below.

Although in the retarded integrals the retardation symbol [ ] usually appears only in the numerators of the integrands, all quantities in the integrals are retarded. In particular, the volume element dV' stands for the retarded volume element $dV'_{ret} = [dV']$ $= d[x']d[y']d[z']$, r stands for the retarded distance $[r]$, and $\mathbf{r}$ stands for the retarded position vector $[\mathbf{r}]$. Note that $[\nabla\rho]$ means "ordinary $\nabla\rho$ used at retarded position," $[\partial\rho/\partial t]$ means "derivative of ordinary ρ with respect to ordinary time used at retarded position," and $[\partial \mathbf{J}/\partial t]$ means "derivative of ordinary **J** with respect to ordinary time used at retarded position."

In the derivations that follow, we shall assume that the point of observation is at $x = y = z = 0$, and we shall only consider a time-independent charge distribution moving with constant velocity in the $-x$ direction. For such a charge distribution, because the charge density is not a function of time, $[\rho] = \rho$, and, because **v** is constant, $[\mathbf{v}] = \mathbf{v}$. Also, as explained in Section 3-1 [see Eqs. (3-1.8), (3-1.3), (3-1.4), (3-1.26), and (3-1.27)], the following relations hold for such a charge distribution

$$[dV'] = \frac{dV'}{1-[\mathbf{r}\cdot\mathbf{v}]/rc}, \tag{5-1.5}$$

$$\frac{\partial \rho}{\partial t} = -\,\mathbf{v}\cdot\nabla'\rho = v\frac{\partial \rho}{\partial x'}, \tag{5-1.6}$$

$$\frac{\partial \mathbf{J}}{\partial t} = -\,(\mathbf{v}\cdot\nabla'\rho)\mathbf{v} = -v^2\frac{\partial \rho}{\partial x'}\mathbf{i}, \tag{5-1.7}$$

$$\begin{aligned}[r]-[\mathbf{r}\cdot\mathbf{v}]/c &= \{x_0'^2+y'^2+z'^2-(y'^2+z'^2)v^2/c^2\}^{1/2}\\ &=\{x_0'^2+(y'^2+z'^2)(1-v^2/c^2)\}^{1/2}=\{x_0'^2+(y'^2+z'^2)/\gamma^2\}^{1/2},\end{aligned} \tag{5-1.8}$$

[we are using the abbreviation $\gamma = (1 - v^2/c^2)^{-1/2}$], and

$$[r]-[\mathbf{r}\cdot\mathbf{v}]/c = r_0\{1-(v^2/c^2)\sin^2\theta\}^{1/2}, \tag{5-1.9}$$

where $\sin^2\theta = (y'^2 + z'^2)/(x_0' + y'^2 + z'^2)$ and θ is the angle between the velocity vector **v** and the vector $[\mathbf{r}_0]$ joining $[dV']$ with the point of observation. For clarity, all retarded quantities and expressions in the above equations are placed between square brackets; the quantities without brackets, and the quantities between braces in Eq. (5-1.8) and (5-1.9) in particular, are present-time quantities. Observe that Eq. (5-1.8) is obtained from Eq. (3-1.26) by replacing y'^2 by $y'^2 + z'^2$; the replacement is needed because we no longer deal with a point charge and therefore cannot assume that the charge is confined to the xy plane.

We can now proceed with the conversion of Eqs. (5-1.1)- (5-1.4). Once again, we shall only consider a time-independent charge distribution moving with constant velocity $\mathbf{v} = -v\mathbf{i}$.

Converting Eq. (5-1.1). Using Eqs. (5-1.5) and (5-1.7) and remembering that ρ and $\mathbf{v}$ are not affected by retardation and that $\nabla\rho$ in Eq. (5-1.1) is the ordinary gradient, we can write Eq. (5-1.1) as

$$\begin{aligned}\mathbf{E} &= -\frac{1}{4\pi\varepsilon_0}\int\frac{\nabla'\rho-(\mathbf{v}\cdot\nabla'\rho)\mathbf{v}/c^2}{[r-\mathbf{r}\cdot\mathbf{v}/c]}dV' \\ &= -\frac{1}{4\pi\varepsilon_0}\int\frac{\nabla'\rho-\mathbf{i}(v^2/c^2)(\partial\rho)/(\partial x')}{[r-\mathbf{r}\cdot\mathbf{v}/c]}dV',\end{aligned} \tag{5-1.10}$$

where only the denominator is retarded. Converting the retarded denominator in Eq. (5-1.10) with the help of Eq. (5-1.8), we obtain the desired equation (we are omitting the subscript "0" at x' for simplicity)

$$\mathbf{E} = -\frac{1}{4\pi\varepsilon_0}\int\frac{\nabla'\rho-\mathbf{i}(v^2/c^2)(\partial\rho)/(\partial x')}{\{x'^2+(y'^2+z'^2)/\gamma^2\}^{1/2}}dV', \tag{5-1.11}$$

where the integral is a "present position" integral, and where all quantities are present-time quantities.

Equation (5-1.11) can be written in an alternative form. Using Eq. (5-1.9) for converting the denominator of the integrand in Eq. (5-1.10), we obtain (omitting the subscript "0" at r for simplicity)

$$\mathbf{E} = -\frac{1}{4\pi\varepsilon_0}\int\frac{\nabla'\rho-\mathbf{i}(v^2/c^2)(\partial/\partial x')\rho}{r\{1-(v^2/c^2)\sin^2\theta\}^{1/2}}dV'. \tag{5-1.12}$$

An even simpler expression for $\mathbf{E}$ of a moving charge distribution can be obtained from Eq. (5-1.3) if the density of the charge under consideration is constant within the volume occupied by the charge. As was shown in Section 2-3, in this case the charge gradient exists only at the surface of the charge, and the volume integral reduces to a surface integral. Equation (5-1.12) becomes then

$$\mathbf{E} = \frac{\rho}{4\pi\varepsilon_0}\oint\frac{d\mathbf{S}' \pm \mathbf{i}(v^2/c^2)dy'dz'}{r\{1-(v^2/c^2)\sin^2\theta\}^{1/2}}, \tag{5-1.13}$$

where the surface element vector $d\mathbf{S}'$ is directed from the charge distribution into the surrounding space, and the sign in front of $\mathbf{i}$ is the same as that of $\partial\rho/\partial x'$.

▼

Example 5-1.1. A thin ring of width w, thickness b, and radius $a \gg b$ carries a uniformly distributed charge q and moves with velocity $\mathbf{v} = -v\mathbf{i}$ along the x axis, which is also the symmetry axis of the ring (Fig. 5.1). Find the electric field produced by the ring at the origin of coordinates when the center of the ring is at a distance x' from the origin.

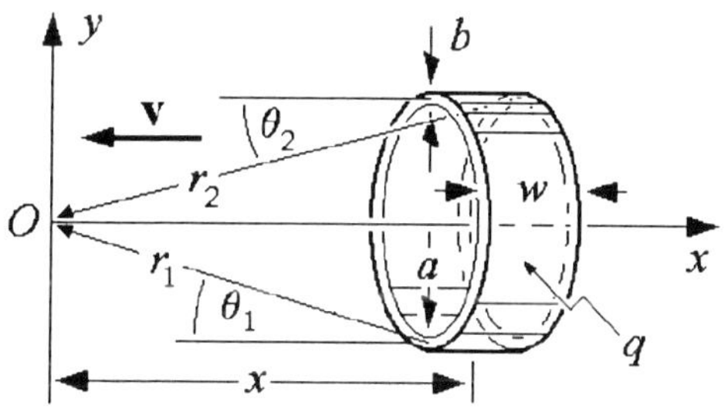

Fig. 5.1 A thin ring of charge q moves with velocity $\mathbf{v} = -v\mathbf{i}$ *along the x axis. Find the electric field at the origin.*

We can solve this problem by using Eq. (5-1.13). By symmetry, only the front (leading) and the back (trailing) surface of the ring contribute to the electric field at the origin. Let the distances from the front and the back surface of the ring to the origin be r_1 and r_2. We then have $r_1 = [(x' - w/2)^2 + a^2]^{1/2}$, $r_2 = [(x' + w/2)^2 + a^2]^{1/2}$, $\sin\theta_1 = a/[(x' - w/2)^2 + a^2]^{1/2}$, $\sin\theta_2 = a/[(x' + w/2)^2 + a^2]^{1/2}$. Equation (5-1.13) becomes therefore

$$\mathbf{E} = \frac{\rho}{4\pi\varepsilon_0}\left(\int \frac{-\{1 - v^2/c^2\}dy'dz'\mathbf{i}}{r_1\{1 - (v^2/c^2)\sin^2\theta_1\}^{1/2}} + \int \frac{\{1 - v^2/c^2\}dy'dz'\mathbf{i}}{r_2\{1 - (v^2/c^2)\sin^2\theta_2\}^{1/2}}\right), \tag{5-1.14}$$

where the integration is over the two flat surfaces of the ring. Substituting the above values for r_1, r_2, $\sin\theta_1$, and $\sin\theta_2$ and taking into account that the area of each flat surface of the ring is $2\pi ab$, we then have

$$\mathbf{E} = \mathbf{i}\frac{\rho(1 - v^2/c^2)2\pi ab}{4\pi\varepsilon_0}\left(\frac{-1}{\{(x' - w/2)^2 + a^2 - v^2a^2/c^2\}^{1/2}} + \frac{1}{\{(x' + w/2)^2 + a^2 - v^2a^2/c^2\}^{1/2}}\right), \tag{5-1.15}$$

or

$$\mathbf{E} = \mathbf{i}\frac{q(1 - v^2/c^2)}{4\pi\varepsilon_0 w}\left(\frac{1}{\{(x' + w/2)^2 + (1 - v^2/c^2)a^2\}^{1/2}} - \frac{1}{\{(x - w/2)^2 + (1 - v^2/c^2)a^2\}^{1/2}}\right). \tag{5-1.16}$$

Example 5-1.2. An infinitely long, thin, straight ribbon of width a and thickness b carries a charge of uniform density ρ and moves along its length with velocity $\mathbf{v} = -v\mathbf{i}$ (Fig. 5.2). The plane of the ribbon is in the xz plane of rectangular coordinates and the center line of the ribbon is on the x axis. Find the electric and magnetic fields produced by the ribbon at the point $P(0, 0, R)$.

We can solve this problem by using Eqs. (5-1.13) and (3-2.5). According to Eq. (5-1.13), the only contribution to the electric field of the ribbon at P comes from the edges of the ribbon located at

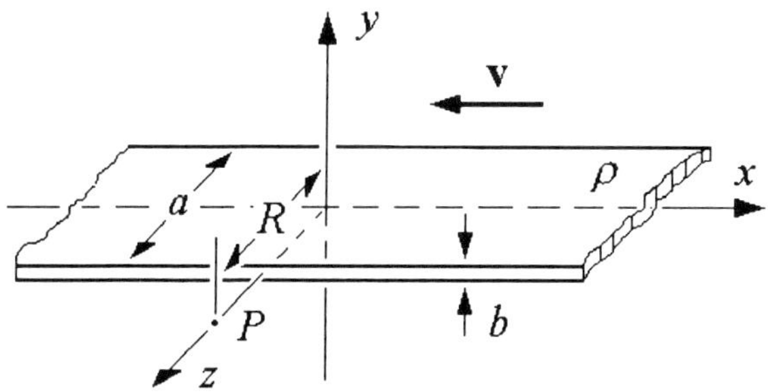

Fig. 5.2 *An infinitely long thin ribbon of charge density ρ moves with uniform velocity* v = − v**i** *along the x axis of rectangular coordinates. Find the electric and magnetic fields produced by the ribbon at the point P.*

$z' = a/2$ and $z' = -a/2$. Let us assume that the ends of the ribbon are at $x' = -L_1$ and $x' = L_2$. By Eqs. (5-1.13), (5-1-9), and (5-1.8), we then have

$$\mathbf{E} = \frac{\rho}{4\pi\varepsilon_0}\left(\int_{-L_1}^{L_2} \frac{\mathbf{k}\,t\,dx'}{\{x'^2+(R-a/2)^2/\gamma^2\}^{1/2}} - \int_{-L_1}^{L_2} \frac{\mathbf{k}\,t\,dx'}{\{x'^2+(R+a/2)^2/\gamma^2\}^{1/2}}\right)$$
$$= \frac{\mathbf{k}\rho b}{4\pi\varepsilon_0}\Big\{\ln\left(x' + \{x'^2+(R-a/2)^2/\gamma^2\}^{1/2}\right) \tag{5-1.17}$$
$$- \ln\left(x' + \{x'^2+(R+a/2)^2/\gamma^2\}^{1/2}\right)\Big\}\Big|_{-L_1}^{L_2},$$

or

$$\mathbf{E} = \frac{\mathbf{k}\rho b}{4\pi\varepsilon_0}\left[\ln\frac{L_2+\{L_2^{\,2}+(R-a/2)^2/\gamma^2\}^{1/2}}{-L_1+\{L_1^{\,2}+(R-a/2)^2/\gamma^2\}^{1/2}} - \ln\frac{L_2+\{L_2^{\,2}+(R+a/2)^2/\gamma^2\}^{1/2}}{-L_1+\{L_1^{\,2}+(R+a/2)^2/\gamma^2\}^{1/2}}\right]. \tag{5-1.18}$$

Since $R - a \ll L_1, L_2$ and $R + a \ll L_1, L_2$, we can expand the expressions in the braces and keep only the leading terms, obtaining

$$\mathbf{E} = \frac{\mathbf{k}\rho b}{4\pi\varepsilon_0}\left[\ln\frac{L_2 + L_2 + (R-a/2)^2/2L_2\gamma^2}{-L_1 + L_1 + (R-a/2)^2/2L_2\gamma^2}\right.$$

$$\left. - \ln\frac{L_2 + L_2 + (R+a/2)^2/2L_2\gamma^2}{-L_1 + L_1 + (R+a/2)^2/2L_2\gamma^2}\right] \qquad (5\text{-}1.19)$$

$$= \frac{\mathbf{k}\rho b}{4\pi\varepsilon_0}\left[\ln\frac{2L_2 + (R-a/2)^2/2L_2\gamma^2}{(R-a/2)^2/2L_1\gamma^2} - \ln\frac{2L_2 + (R+a/2)^2/2L_2\gamma^2}{(R+a/2)^2/2L_1\gamma^2}\right]$$

and, finally,

$$\mathbf{E} = \mathbf{k}\frac{\rho b}{2\pi\varepsilon_0}\ln\frac{(R+a/2)}{(R-a/2)}. \qquad (5\text{-}1.20)$$

To find the magnetic field, we will use Eq. (3-2.5). By Eqs. (3-2.5) and (5-1.20), we have

$$\mathbf{H} = \varepsilon_0\mathbf{v}\times\mathbf{E} = (-\mathbf{i}\times\mathbf{k})\frac{\rho v b}{2\pi}\ln\frac{(R+a/2)}{(R-a/2)} \qquad (5\text{-}1.21)$$

or

$$\mathbf{H} = \mathbf{j}\frac{Jb}{2\pi}\ln\frac{(R+a/2)}{(R-a/2)} = \mathbf{j}\frac{I}{2\pi a}\ln\frac{(R+a/2)}{(R-a/2)}, \qquad (5\text{-}1.22)$$

where J is the current density and I is the current formed by the ribbon.

Observe that Eq. (5-1.22) is the same as that obtained for this current configuration by means of Biot-Savart's law (or its equivalent),[1] which, taking into account the diversity and complexity of the theoretical considerations leading to Eqs. (5-1.13) and (3-2.5), and observing that Eqs. (5-1.13) and (3-2.5) appear to have no connection with Biot-Savart's law, is quite remarkable.

▲

Converting Eq. (5-1.3). As before, we assume that the charge is time independent and moves with constant velocity $\mathbf{v} = -v\mathbf{i}$. Using Eqs. (5-1.6) and (5-1.7), we can write Eq. (5-1.3) as

$$\mathbf{E} = \frac{1}{4\pi\varepsilon_0}\int \frac{[\rho]}{r^3}\mathbf{r}\,dV' + \frac{1}{4\pi\varepsilon_0 c}\int \frac{-[\mathbf{v}\cdot\nabla'\rho]\mathbf{r} + \mathbf{v}[\mathbf{v}\cdot\nabla'\rho]r/c}{r^2}dV'. \tag{5-1.23}$$

Note that $\nabla'\rho$ in this equation represents the ordinary gradient, that is, the gradient with respect to the ordinary source-point coordinates. For the calculations that follow, we need to convert $\nabla'\rho$ into the gradient with respect to the *retarded* coordinates. According to Eq. (3-1.7),

$$d[x'] = \frac{dx'}{1-[\mathbf{r}\cdot\mathbf{v}]/[r]c}, \tag{5-1.24}$$

and therefore

$$\frac{\partial}{\partial x'} = \frac{1}{1-[\mathbf{r}\cdot\mathbf{v}]/[r]c}\frac{\partial}{\partial[x']}. \tag{5-1.25}$$

Since $\mathbf{v}$ is along the x axis, the y' and z' are not affected by retardation, so that $\partial/\partial y' = \partial/\partial[y']$ and $\partial/\partial z' = \partial/\partial[z']$. Hence

$$[\mathbf{v}\cdot\nabla'\rho] = \frac{[\mathbf{v}]\cdot[\nabla'][\rho]}{1-[\mathbf{r}\cdot\mathbf{v}]/[r]/c}. \tag{5-1.26}$$

Substituting this expression into Eq. (5-1.23), we obtain

$$\mathbf{E} = \frac{1}{4\pi\varepsilon_0}\int_{ret} \frac{\rho}{r^3}\mathbf{r}\,dV' + \frac{1}{4\pi\varepsilon_0 c}\int_{ret} \frac{(\mathbf{v}r/c-\mathbf{r})\mathbf{v}\cdot\nabla'\rho}{r^2(1-\mathbf{v}\cdot\mathbf{r}/rc)}dV', \tag{5-1.27}$$

where all the quantities under the integral signs are retarded, and where we have replaced the retardation brackets in the integrands by the subscript "*ret*" at the integral signs.

Let us designate the last term in Eq. (5-1.27) as $\mathbf{E}_2$. We have

$$\mathbf{E}_2 = \frac{1}{4\pi\varepsilon_0 c}\int_{ret} \frac{(\mathbf{v}r/c-\mathbf{r})\mathbf{v}\cdot\nabla'\rho}{r^2(1-\mathbf{v}\cdot\mathbf{r}/rc)}dV'. \tag{5-1.28}$$

To convert this integral to the present position of the charge, we shall first eliminate $\nabla'\rho$ from it. To do so, we shall write Eq. (5-1.28) in terms of its Cartesian components. For the x component we have, remembering that $\mathbf{v} = -v\mathbf{i}$ and that $\mathbf{r} = -(x'\mathbf{i} + y'\mathbf{j} + z'\mathbf{k})$,

$$E_{2x} = -\frac{1}{4\pi\varepsilon_0 c}\int_{ret} \frac{(vr/c - x')\mathbf{v}\cdot\nabla'\rho}{r^2(1-\mathbf{v}\cdot\mathbf{r}/rc)}dV'. \qquad (5\text{-}1.29)$$

Let us now factor out $\mathbf{v}\cdot$ and let us write the integral as a difference of two integrals,

$$E_{2x} = -\frac{\mathbf{v}\cdot}{4\pi\varepsilon_0 c}\int_{ret} \frac{(vr/c - x')\nabla'\rho}{r^2(1-\mathbf{v}\cdot\mathbf{r}/rc)}dV' \qquad (5\text{-}1.30)$$
$$= -\frac{\mathbf{v}\cdot}{4\pi\varepsilon_0 c}\left\{\int_{ret}\nabla'\frac{(vr/c - x')\rho}{r^2(1-\mathbf{v}\cdot\mathbf{r}/rc)}dV' - \int_{ret}\rho\nabla'\frac{(vr/c - x')}{r^2(1-\mathbf{v}\cdot\mathbf{r}/rc)}dV'\right\}.$$

The first integral in the last expression can be converted into a surface integral by means of Gauss's theorem of vector analysis [vector identity (V-19)], and since there is no charge outside the charge distribution under consideration, the integral vanishes. Differentiating the integrand in the second integral, collecting terms, reintroducing $\mathbf{v}\cdot$ under the integral sign, and simplifying, we obtain

$$E_{2x} = \frac{1}{4\pi\varepsilon_0}\int_{ret}\rho\frac{\{v^2/c^2 - 2\mathbf{v}\cdot\mathbf{r}/rc + (\mathbf{v}\cdot\mathbf{r}/rc)^2\}x' - (v^2/c^2 - 1)vr/c}{r^3(1-\mathbf{v}\cdot\mathbf{r}/rc)^2}dV'. \qquad (5\text{-}1.31)$$

Proceeding in the same manner with the y and z components of Eq. (5-1.28), we obtain

$$E_{2y} = \frac{1}{4\pi\varepsilon_0}\int_{ret}\rho\frac{\{v^2/c^2 - 2\mathbf{v}\cdot\mathbf{r}/rc + (\mathbf{v}\cdot\mathbf{r}/rc)^2\}y'}{r^3(1-\mathbf{v}\cdot\mathbf{r}/rc)^2}dV' \qquad (5\text{-}1.32)$$

and

$$E_{2z} = \frac{1}{4\pi\varepsilon_0}\int_{ret} \rho \frac{\{v^2/c^2 - 2\mathbf{v}\cdot\mathbf{r}/rc + (\mathbf{v}\cdot\mathbf{r}/rc)^2\}z'}{r^3(1-\mathbf{v}\cdot\mathbf{r}/rc)^2} dV' . \quad (5\text{-}1.33)$$

Multiplying Eqs. (5-1.31)-(5-1.33), respectively, by **i**, **j**, and **k** and then adding them together, we again obtain a single vector equation for $\mathbf{E}_2$:

$$\mathbf{E}_2 = \frac{1}{4\pi\varepsilon_0}\int_{ret} \rho \frac{\{2\mathbf{v}\cdot\mathbf{r}/rc - (\mathbf{v}\cdot\mathbf{r}/rc)^2 - v^2/c^2\}\mathbf{r} + (v^2/c^2 - 1)\mathbf{v}r/c}{r^3(1-\mathbf{v}\cdot\mathbf{r}/rc)^2} dV'. \quad (5\text{-}1.34)$$

Let us now rewrite Eq. (5-1.27) using Eq. (5-1.34) for the second integral of Eq. (5-1.27). We then have

$$\mathbf{E} = \frac{1}{4\pi\varepsilon_0}\int_{ret} \frac{\rho}{r^3}\mathbf{r}\, dV' \quad (5\text{-}1.35)$$
$$+ \frac{1}{4\pi\varepsilon_0}\int_{ret} \rho \frac{\{2\mathbf{v}\cdot\mathbf{r}/rc - (\mathbf{v}\cdot\mathbf{r}/rc)^2 - v^2/c^2\}\mathbf{r} + (v^2/c^2 - 1)\mathbf{v}r/c}{r^3(1-\mathbf{v}\cdot\mathbf{r}/rc)^2} dV'.$$

Adding the two integrals, we obtain

$$\mathbf{E} = \frac{1}{4\pi\varepsilon_0}\int_{ret} \rho \frac{(1 - v^2/c^2)(\mathbf{r} - \mathbf{v}r/c)}{r^3(1-\mathbf{v}\cdot\mathbf{r}/rc)^2} dV' . \quad (5\text{-}1.36)$$

We shall now convert the retarded integral in Eq. (5-1.36) to the present position of the charge. Replacing the retarded dV' in Eq. (5-1.36) by ordinary dV' with the help of Eq. (5-1.5) and writing $1/\gamma^2$ for $1 - v^2/c^2$, we have

$$\mathbf{E} = \frac{1}{4\pi\varepsilon_0\gamma^2}\int \frac{\rho([\mathbf{r}] - \mathbf{v}[r]/c)}{[r]^3(1-\mathbf{v}\cdot[\mathbf{r}]/[r]c)^3} dV' , \quad (5\text{-}1.37)$$

where, since ρ, **v**, v, and c do not depend on time, only **r** and r are retarded. But according to Eq. (3-1.19), the present-position vector $\mathbf{r}_0$ and the retarded position vector $\mathbf{r}$ are connected by the relation

$$\mathbf{r}_0 = [\mathbf{r}] - \mathbf{v}[r]/c, \tag{5-1.38}$$

so that the numerator in Eq. (5-1.37) is simply the present-position vector $\mathbf{r}_0$. Furthermore, according to Eq. (5-1.9), the denominator is simply

$$r_0^3\{1-(v^2/c^2)\sin^2\theta\}^{3/2}, \tag{5-1.39}$$

where r_0 is the distance from the present-position volume element dV' to the point of observation, and θ is the angle between $\mathbf{v}$ and $\mathbf{r}_0$. Hence Eq. (5-1.37) can be written as

$$\mathbf{E} = \frac{1}{4\pi\varepsilon_0\gamma^2}\int\frac{\rho\mathbf{r}_0}{r_0^3\{1-(v^2/c^2)\sin^2\theta\}^{3/2}}dV', \tag{5-1.40}$$

where the integration is over the volume of the charge at its present position.

▼

Example 5-1.3. An irregularly shaped electric charge distribution of total charge q moves with constant velocity $\mathbf{v} = v\mathbf{i}$. The longest linear dimension of the charge distribution is a. Find the electric field produced by the charge at a distance $r \gg a$ from the charge.

We can solve the problem by using Eq. (5-1.40). Since $r \gg a$, we can assume r and θ to be the same for all points of the charge. Therefore we can factor out $\mathbf{r}$ and the denominator of the integrand in Eq. (5-1.40), obtaining

$$\begin{aligned}\mathbf{E} &= \frac{\mathbf{r}_0}{4\pi\varepsilon_0\gamma^2 r_0^3[1-(v^2/c^2)\sin^2\theta]^{3/2}}\int\rho dV' \\ &= \frac{q\mathbf{r}_0}{4\pi\varepsilon_0\gamma^2 r_0^3[1-(v^2/c^2)\sin^2\theta]^{3/2}}.\end{aligned} \tag{5-1.41}$$

▲

Converting Eqs. (5-1.2) and (5-1.4). The retarded integrals for the magnetic fields in Eq. (5-1.2) and (5-1.4) can be converted to the present position of the charge in the same manner as the integrals in Eqs. (5-1.1) and (5-1.3) for the electric field. However, there is no need to resort to this conversion process, because by Eq. (3-2.6) the electric and magnetic fields of any uniformly moving charge distribution are connected by the relation

$$\mathbf{B} = (\mathbf{v} \times \mathbf{E})/c^2. \tag{5-1.42}$$

From Eqs. (5-1.12) and (5-1.42) we then have, noting that $\mathbf{v} \times \mathbf{i} = 0$,

$$\mathbf{B} = -\frac{1}{4\pi\varepsilon_0 c^2}\int \frac{\mathbf{v} \times \nabla'\rho}{r_0\{1-(v^2/c^2)\sin^2\theta\}^{1/2}}\,dV'. \tag{5-1.43}$$

From Eqs. (5-1.13) and (5-1.42) we have

$$\mathbf{B} = \frac{\rho}{4\pi\varepsilon_0 c^2}\oint \frac{\mathbf{v} \times d\mathbf{S}'}{r_0\{1-(v^2/c^2)\sin^2\theta\}^{1/2}}\,dV'. \tag{5-1.44}$$

And from Eqs. (5-1.40) and (5-1.42) we have

$$\mathbf{B} = \frac{1}{4\pi\varepsilon_0 c^2\gamma^2}\int \frac{\rho\mathbf{v} \times \mathbf{r}_0}{r_0^3\{1-(v^2/c^2)\sin^2\theta\}^{3/2}}\,dV'. \tag{5-1.45}$$

5-2. Converting Retarded Potential Integrals for Uniformly Moving Charge Distributions into Present-Time (Present Position) Integrals

We know from Chapter 2, Eqs. (2-4.5) and (2-4.2), that the electric potential φ and the magnetic vector potential $\mathbf{A}$ of time-variable charge and current distributions in a vacuum can be found from the retarded integrals

$$\varphi = \frac{1}{4\pi\varepsilon_0}\int\frac{[\rho]}{r}dV' \tag{5-2.1}$$

and

$$\mathbf{A} = \frac{\mu_0}{4\pi}\int\frac{[\mathbf{J}]}{r}dV'. \tag{5-2.2}$$

As we shall presently see, for time-independent charge distributions moving with constant velocity, these integrals can be converted to the "present" position of the charge, so that the integration is performed not over the retarded volume, but over the volume that the charge distribution occupies at the moment t for which the potentials are being determined.[2]

Converting Eq. (5-2.1). Using Eq. (5-1.5) and remembering that ρ and $\mathbf{v}$ are not affected by retardation, we can write Eq. (5-2.1) as

$$\varphi = \frac{1}{4\pi\varepsilon_0}\int\frac{\rho}{[r-\mathbf{r}\cdot\mathbf{v}/c]}dV', \tag{5-2.3}$$

where only the denominator is retarded. Converting the retarded denominator in Eq. (5-2.3) with the help of Eq. (5-1.8), we obtain the desired equation

$$\varphi = \frac{1}{4\pi\varepsilon_0}\int\frac{\rho}{\{x'^2+(y'^2+z'^2)/\gamma^2\}^{1/2}}dV', \tag{5-2.4}$$

where the integral is a "present position" integral, and where all quantities are present-time quantities.

Equation (5-2.4) can be written in an alternative form. Using Eqs. (5-1.8) and (5-1.9) for converting the denominator of the integrand in Eq. (5-2.4), we obtain

$$\varphi = \frac{1}{4\pi\varepsilon_0}\int\frac{\rho}{r\{1-(v^2/c^2)\sin^2\theta\}^{1/2}}dV'. \tag{5-2.5}$$

Equations (5-2.4) and (5-2.5) can be further modified so that the potential is expressed not in terms of the charge density ρ as such, but in terms of $\nabla\rho$ (that is, in terms of the "charge inhomogeneities"). This can be done as follows.[3]

Taking into account that the position vector $\mathbf{r}$ is directed toward the point of observation, so that $\mathbf{r} = -x'\mathbf{i} - y'\mathbf{j} - z'\mathbf{k}$ and $\nabla'\cdot\mathbf{r} = -3$, we write

$$\begin{aligned}\nabla'\cdot\frac{\mathbf{r}\rho}{\{x'^2+(y'^2+z'^2)/\gamma^2\}^{1/2}} &= \frac{\mathbf{r}}{\{x'^2+(y'^2+z'^2)/\gamma^2\}^{1/2}}\cdot\nabla'\rho \\ &+\rho\nabla'\cdot\frac{\mathbf{r}}{\{x'^2+(y'^2+z'^2)/\gamma^2\}^{1/2}} \\ &= \frac{\mathbf{r}}{\{x'^2+(y'^2+z'^2)/\gamma^2\}^{1/2}}\cdot\nabla'\rho - \frac{3\rho}{\{x'^2+(y'^2+z'^2)/\gamma^2\}^{1/2}} \\ &- \frac{\mathbf{r}\cdot\{x'\mathbf{i}+(y'\mathbf{j}+z'\mathbf{k})/\gamma^2\}\rho}{\{x'^2+(y'^2+z'^2)/\gamma^2\}^{3/2}} \\ &= \frac{\mathbf{r}\cdot\nabla'\rho}{\{x'^2+(y'^2+z'^2)/\gamma^2\}^{1/2}} - \frac{2\rho}{\{x'^2+(y'^2+z'^2)/\gamma^2\}^{1/2}}.\end{aligned} \tag{5-2.6}$$

Using Eq. (5-2.6) and Eq. (5-2.4), we can now express the potential as

$$\begin{aligned}\varphi = &-\frac{1}{8\pi\varepsilon_0}\int\nabla'\cdot\frac{\mathbf{r}\rho}{\{x'^2+(y'^2+z'^2)/\gamma^2\}^{1/2}}\,dV' \\ &+\frac{1}{8\pi\varepsilon_0}\int\frac{\mathbf{r}\cdot\nabla\rho}{\{x'^2+(y'^2+z'^2)/\gamma^2\}^{1/2}}\,dV'.\end{aligned} \tag{5-2.7}$$

The first integral in this equation can be transformed into a surface integral over all space by means of Gauss's theorem of vector analysis [vector identity (V-19)], and, since there are no charges at infinity, the integral vanishes. Hence the potential can be written as

$$\varphi = \frac{1}{8\pi\varepsilon_0}\int \frac{\mathbf{r}\cdot\nabla'\rho}{\{x'^2+(y'^2+z'^2)/\gamma^2\}^{1/2}}\,dV', \tag{5-2.8}$$

or, by using Eqs. (5-1.8) and (5-1.9), as

$$\varphi = \frac{1}{8\pi\varepsilon_0}\int \frac{\mathbf{r}\cdot\nabla'\rho}{r\{1-(v^2/c^2)\sin^2\theta\}^{1/2}}\,dV'. \tag{5-2.9}$$

Equations (5-2.8) and (5-2.9) can be written in a much simpler form, if ρ is constant within the charge distribution. In this case $\nabla'\rho$ is different from zero only in the surface layer of the charge distribution, where the charge changes from ρ within the distribution to zero outside the distribution. We then have $\nabla'\rho = (\rho/\tau)\mathbf{n}_u$, where τ is the thickness of the surface layer of the distribution, and $\mathbf{n}_u$ is a unit vector normal to the surface of the distribution and directed into the distribution. The volume element dV' in Eqs. (5-2.8) and (5-2.9) becomes then $\tau dS'$, where dS' is a surface area element of the distribution, and therefore Eqs. (5-2.8) and (5-2.9) reduce to

$$\varphi = -\frac{\rho}{8\pi\varepsilon_0}\oint \frac{\mathbf{r}\cdot d\mathbf{S}'_{out}}{\{x'^2+(y'^2+z'^2)/\gamma^2\}^{1/2}}, \tag{5-2.10}$$

and

$$\varphi = -\frac{\rho}{8\pi\varepsilon_0}\oint \frac{\mathbf{r}\cdot d\mathbf{S}'_{out}}{r\{1-(v^2/c^2)\sin^2\theta\}^{1/2}}, \tag{5-2.11}$$

where $d\mathbf{S}_{out}$ is a surface element vector directed from the charge distribution into the surrounding space.

Converting Eq. (5-2.2). The current density produced by a uniformly moving charge distribution is $\mathbf{J} = \rho\mathbf{v}$ with $\mathbf{v}$ = const. Since $\mu_0\varepsilon_0 = 1/c^2$, the vector potential $\mathbf{A}$ for such a charge distribution is, by Eqs. (5-2.2) and (5-2.1), $\mathbf{A} = \mathbf{v}\varphi/c^2$. Hence, using Eqs. (5-2.5), (5-2.9), and (5-2.11), we have

$$\mathbf{A} = \frac{\mathbf{v}\varphi}{c^2}, \tag{5-2.12}$$

$$\mathbf{A} = \frac{\mathbf{v}}{4\pi\varepsilon_0 c^2}\int \frac{\rho}{r\{1-(v^2/c^2)\sin^2\theta\}^{1/2}}\,dV', \tag{5-2.13}$$

$$\mathbf{A} = \frac{\mathbf{v}}{8\pi\varepsilon_0 c^2}\int \frac{\mathbf{r}\cdot\nabla'\rho}{r\{1-(v^2/c^2)\sin^2\theta\}^{1/2}}\,dV', \tag{5-2.14}$$

and

$$\mathbf{A} = -\frac{\mathbf{v}}{8\pi\varepsilon_0 c^2}\oint \frac{\rho\,\mathbf{r}\cdot d\mathbf{S}'_{out}}{r\{1-(v^2/c^2)\sin^2\theta\}^{1/2}}, \tag{5-2.15}$$

and similar expressions corresponding to Eqs. (5-2.4), (5-2.8), and (5-2.10):

$$\mathbf{A} = \frac{\mathbf{v}}{4\pi\varepsilon_0 c^2}\int \frac{\rho}{\{x'^2+(y'^2+z'^2)/\gamma^2\}^{1/2}}\,dV', \tag{5-2.16}$$

$$\mathbf{A} = \frac{\mathbf{v}}{8\pi\varepsilon_0 c^2}\int \frac{\mathbf{r}\cdot\nabla'\rho}{\{x'^2+(y'^2+z'^2)/\gamma^2\}^{1/2}}\,dV', \tag{5-2.17}$$

$$\mathbf{A} = -\frac{\mathbf{v}\rho}{8\pi\varepsilon_0 c^2}\oint \frac{\mathbf{r}\cdot d\mathbf{S}'_{out}}{\{x'^2+(y'^2+z'^2)/\gamma^2\}^{1/2}}. \tag{5-2.18}$$

▼

Example 5-2.1. An irregularly shaped electric charge distribution of total charge q moves with constant velocity $\mathbf{v} = -v\mathbf{i}$. The longest linear dimension of the charge distribution is a. Find the electric and magnetic potentials produced by the charge at a distance $r \gg a$ from the charge.

We can solve the problem by using Eqs. (5-2.5) and (5-2.13). Since $r \gg a$, we can assume r and θ to be the same for all points of the charge. Therefore we can factor out the denominator of the integrands in Eq. (5-2.5) and (5-2.13), obtaining

$$\varphi = \frac{q}{4\pi\varepsilon_0 r\{1-(v^2/c^2)\sin^2\theta\}^{1/2}}, \tag{5-2.19}$$

$$\mathbf{A} = \frac{\mathbf{v}q}{4\pi\varepsilon_0 c^2 r\{1-(v^2/c^2)\sin^2\theta\}^{1/2}}. \tag{5-2.20}$$

▲

5-3. Some Peculiarities of the Expressions for the Fields and Potentials Derived in this Chapter

Three peculiarities of the equations for the electric and magnetic fields and potentials derived in this chapter should be noted.

First, in the equations developed in the preceding chapters we used both retarded and present-time (present position) coordinates, and therefore we needed to use different notation for the two types of coordinates. In particular, we designated the present position vector as $\mathbf{r}_0$ and the x component of this vector as x_0', while we designated the retarded position vector as $\mathbf{r}$ and its x component as x'. However, since all the resulting expressions for the fields and potentials developed in this chapter are for the present position of the charge distributions, there is no longer a need to use the subscript "0" at $\mathbf{r}$ or x'. Therefore, in the field and potential equations obtained in this chapter $\mathbf{r}$ and x' stand for the present-time (present position) coordinates.

Second, in deriving our equations for the potentials of moving charge distributions, we assumed that the field point (the point for which the potentials are determined) was at the origin. However, in practical application of the potentials it is usually necessary to differentiate the potentials with respect to the field point. In particular, for finding electric and magnetic fields from potentials it is necessary to operate upon the electric and magnetic potentials with the operator ∇ (which operates upon the field point coordinates). Therefore, in general, the field point must be allowed to vary.

We can easily convert our equations for the potentials (and fields) into equations with a variable field point. Let us designate the coordinates of this point as x, y, and z. If we then replace the x', y', and z' coordinates appearing explicitly or implicitly in our equations for potentials or fields by $(x - x')$, $(y - y')$, and $(z - z')$, respectively, the new equations will apply to fields and potentials determined for the field point x, y, z. However, if the charge density ρ within the charge distribution under consideration is constant, we can differentiate the potentials with respect to the field point without actually replacing the x', y', z' coordinates at all, because in this case, by vector identity (V-27), the only difference between the differentiation of the integrands with respect to x', y', z' and with respect to x, y, z is in the sign of the resulting expression. Thus, in the case of constant ρ, we can compute electric and magnetic fields from the potentials derived in this chapter without changing the coordinates, provided that after placing ∇ under the integral sign we replace it by $-\nabla'$ (see Example 5-3.1).

Third, all the fields and potentials derived in this chapter are "snapshots" representing only the instantaneous values of the observed fields and potentials. In reality the fields and potentials of a moving charge distribution vary as the charge distribution moves relative to the point of observation. For practical applications it may be necessary to determine time derivatives of the fields and potentials. Therefore, in general, the fields and potentials must be expressed as a function of time. This can be easily done by noting that as a charge distribution moves (with constant speed), the present position of dV' (or $d\mathbf{S}'$) is given by $x' \mp vt$ (the minus applies to motion against the x axis, the plus applies to the motion in the direction of the x axis). Thus all we need to do for introducing the time dependence into the fields and potentials derived in this chapter is to replace x' appearing explicitly or implicitly in our field and potential equations by $x' \mp vt$ (see Example 5-3.1, see also Examples 4-1.2 and 4-5.1).

▼

Example 5-3.1 A very long hollow cylinder of wall thickness b and radius $a \gg b$ carries a uniformly distributed charge of density ρ and moves with velocity $\mathbf{v} = -v\mathbf{i}$ along the x axis, which is also the symmetry axis of the cylinder (Fig. 5.3). Find the electric field produced by the cylinder at the origin of coordinates when the leading end of the cylinder is at a distance x' from the origin.

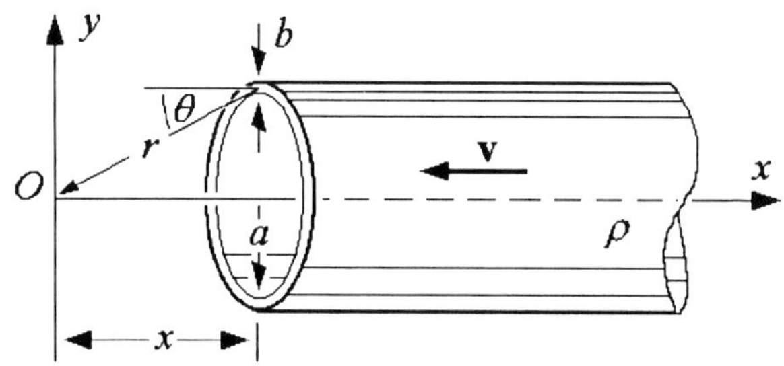

Fig. 5.3 A very long cylinder of charge density ρ moves with uniform velocity $\mathbf{v} = -v\mathbf{i}$ *along the x axis. Find the electric field produced by the cylinder at the origin.*

We shall solve this problem by using Eqs. (5-2.4) and (5-2.16). Applying the relation $\mathbf{E} = -\nabla\varphi - \partial\mathbf{A}/\partial t$ [this is Eq. (2-4.8) derived in Section 2.4] to Eqs. (5-2.4) and (5-2.16), we obtain

$$\mathbf{E} = -\nabla\left(\frac{1}{4\pi\varepsilon_0}\int\frac{\rho}{\{x'^2+(y'^2+z'^2)/\gamma^2\}^{1/2}}dV'\right)$$
$$-\frac{\partial}{\partial t}\left(\frac{\mathbf{v}}{4\pi\varepsilon_0 c^2}\int\frac{\rho}{\{x'^2+(y'^2+z'^2)/\gamma^2\}^{1/2}}dV'\right). \tag{5-3.1}$$

In Eq. (5-3.1), ∇ operates upon the field point coordinates x, y, z, which do not appear in Eq. (5-3.1). However, as explained above, for constant ρ we can leave the first integral in Eq. (5-3.1) as it now is, provided that for the actual differentiation we replace ∇ by $-\nabla'$. Placing ∇ under the integral sign and replacing it by $-\nabla'$, we have for the part of the electric field due to φ (using $\mathbf{E} = \mathbf{E}_\varphi + \mathbf{E}_A$)

$$\mathbf{E}_\varphi = \frac{1}{4\pi\varepsilon_0}\int\nabla'\frac{\rho}{\{x'^2+(y'^2+z'^2)/\gamma^2\}^{1/2}}dV'. \tag{5-3.2}$$

To differentiate the second integral in Eq. (5-3.1), we must first express the integrand as a function of t. Replacing x' in the integrand by $x' - vt$, placing $\partial/\partial t$ under the integral sign, and differentiating the integrand, we then have for the part of the electric field due to $\mathbf{A}$

$$\mathbf{E}_A = -\frac{\mathbf{v}}{4\pi\varepsilon_0 c^2}\int\frac{\rho(x'-vt)v}{\{(x'-vt)^2+(y'^2+z'^2)/\gamma^2\}^{3/2}}dV', \tag{5-3.3}$$

or, setting $t = 0$,

$$\mathbf{E}_A = -\frac{\mathbf{v}}{4\pi\varepsilon_0 c^2}\int\frac{vx'\rho}{\{x'^2+(y'^2+z'^2)/\gamma^2\}^{3/2}}dV', \tag{5-3.4}$$

which, as one can easily verify by direct differentiation, is the same as

$$\mathbf{E}_A = -\frac{\mathbf{v}}{4\pi\varepsilon_0 c^2}\mathbf{v}\cdot\int\nabla'\frac{\rho}{\{x'^2+(y'^2+z'^2)/\gamma^2\}^{1/2}}dV'. \tag{5-3.5}$$

The total field is therefore

$$\begin{aligned}\mathbf{E} = & \frac{1}{4\pi\varepsilon_0}\int \nabla' \frac{\rho}{\{x'^2 + (y'^2 + z'^2)/\gamma^2\}^{1/2}} dV' \\ & - \frac{\mathbf{v}}{4\pi\varepsilon_0 c^2}\mathbf{v}\cdot\int \nabla' \frac{\rho}{\{x'^2 + (y'^2 + z'^2)/\gamma^2\}^{1/2}} dV'.\end{aligned} \tag{5-3.6}$$

Using now Gauss's theorem of vector analysis [vector identity (V-19)], we can convert the two integrals into integrals over the surface of the cylinder, obtaining[4]

$$\begin{aligned}\mathbf{E} = & \frac{\rho}{4\pi\varepsilon_0}\left\{\oint \frac{d\mathbf{S}_{out}}{\{x'^2 + (y'^2 + z'^2)/\gamma^2\}^{1/2}}\right. \\ & \left. - \frac{\mathbf{v}}{c^2}\mathbf{v}\cdot\oint \frac{d\mathbf{S}_{out}}{\{x'^2 + (y'^2 + z'^2)/\gamma^2\}^{1/2}}\right\},\end{aligned} \tag{5-3.7}$$

where $d\mathbf{S}_{out}$ is a surface element vector directed outward from the volume of the cylinder.

By the symmetry of the system, the electric field at the point of observation has only the x component. The only surfaces of the cylinder contributing to that component are the surfaces of the leading and trailing ends of the cylinder. However, since the cylinder is very long, the contribution of the trailing end is negligible. Furthermore, since the cylinder's wall is thin, the integration over the leading end can be replaced by the multiplication of the integrand by the surface area $S = 2\pi ab$ of the leading end's wall. Taking into account that $\mathbf{v} = -v\mathbf{i}$, that for the leading end $y'^2 + z'^2 = a^2$, $d\mathbf{S}_{\text{out}} = -dS\mathbf{i}$, and $\mathbf{v}\cdot d\mathbf{S}_{\text{out}} = vdS$, we finally obtain for the "snapshot" of the electric field produced by the cylinder at the point of observation

$$\mathbf{E} = -\frac{\rho ab(1 - v^2/c^2)}{2\varepsilon_0\{x'^2 + a^2(1 - v^2/c^2)\}^{1/2}}\mathbf{i}. \tag{5-3.8}$$

Example 5-3.2 A line charge of length $2L$ and linear charge density λ moves with constant velocity $\mathbf{v} = -v\mathbf{i}$ in the xy plane of a rectangular system of coordinates at a distance $y = R$ above the x axis. The point of observation is at the origin. Find the electric potential, the electric field, and the magnetic field at the origin at the moment when the two ends of the charge are at equal distances L from the y axis and then obtain the limiting value of the fields for a very long charge.

To find the electric potential, we use Eq. (5-2.4) with $\rho dV'$ replaced by $\lambda dx'$. Integrating over the length of the line charge we then have

$$\begin{aligned} \varphi &= \frac{1}{4\pi\varepsilon_0}\int_{-L/2}^{L/2}\frac{\lambda}{(x'^2+y'^2/\gamma^2)^{1/2}}dx' \\ &= \frac{\lambda}{4\pi\varepsilon_0}\ln\{x'+(x'^2+y'^2/\gamma^2)^{1/2}\}\Big|_{-L/2}^{L/2}, \end{aligned} \tag{5-3.9}$$

or

$$\varphi = \frac{\lambda}{4\pi\varepsilon_0}\ln\frac{\{L+(L^2+y'^2/\gamma^2)^{1/2}\}}{\{-L+(L^2+y'^2/\gamma^2)^{1/2}\}}. \tag{5-3.10}$$

To find the electric field, we differentiate the potential with respect to y', using the *positive* derivative. The result is

$$\mathbf{E} = -\frac{\lambda}{2\pi\varepsilon_0 y'(1+y'^2/\gamma^2L^2)^{1/2}}\mathbf{j} = -\frac{\lambda}{2\pi\varepsilon_0 R(1+R^2/\gamma^2L^2)^{1/2}}\mathbf{j}. \tag{5-3.11}$$

The magnetic field of the line charge is, by Eqs. (5-3.11) and (3-2.5),

$$\mathbf{H} = \frac{\lambda v}{2\pi R(1+R^2/\gamma^2L^2)^{1/2}}\mathbf{k}. \tag{5-3.12}$$

For a very long charge, $L \gg R$, so that Eqs. (5-3.11) and (5-3.12) reduce to

$$\mathbf{E} = -\frac{\lambda}{2\pi\varepsilon_0 R}\mathbf{j} \tag{5-3.13}$$

and

$$\mathbf{H} = \frac{\lambda v}{2\pi R}\mathbf{k}. \tag{5-3.14}$$

It is interesting to note that the electric field given by Eq. (5-3.13) is the same as that of a stationary infinitely long line charge, and that the magnetic field given by Eq. (5-3.14) is the same as the magnetic field produced by a current $I = \lambda v$.[5]

▲

References and Remarks for Chapter 5

1. See Oleg D. Jefimenko, *Electricity and Magnetism*, 2nd ed., (Electret Scientific, Star City, 1989) pp. 354 - 356.
2. For somewhat different derivations see Oleg D. Jefimenko, "Retardation and relativity: new integrals for electric and magnetic potentials of time-independent charge distributions moving with constant velocity," Eur. J. Phys. **17**, 257-264 (1996).
3. For an analogous transformation of an integral for the electric field of a stationary charge see Ref. 1, pp. 92-94 and 101-103.
4. The surface of integration remains *within* the charge distribution, just touching the surface layer of the charge, but not stepping out of the charge distribution into the space where there is no charge. See Section 2-3.
5. See, for example, Ref. 1, pp. 98, 99 and 330-332.

II

RELATIVITY

6

FROM ELECTROMAGNETIC RETARDATION TO RELATIVITY

In the preceding chapters we saw how electric and magnetic fields and potentials of moving charge distributions could be determined on the basis of the theory of electromagnetic retardation. In this and in several chapters that follow we shall acquaint ourselves with an alternative method of determining the fields and potentials of moving charge distributions. This alternative method is based on the principle of relativity and its application to electromagnetic phenomena.

6-1. Relativistic Electromagnetism, Relativistic Terminology, the Principle of Relativity, and Theories of Relativity

We shall enter now into the domain of *relativistic electromagnetism*. The theory of relativistic electromagnetism makes use of some special words and expressions a clear understanding of which is imperative for a proper understanding and use of the theory. A frequently used word in that theory is the *laboratory*. The laboratory is simply a place where instruments and devices for measuring and observing physical phenomena are located. Unless otherwise stated, the laboratory is assumed to be stationary. Another frequently used expression is the *frame of reference*. Physically, a frame of reference is the same as the laboratory. However, a frame of reference can be stationary as well as moving and is depicted graphically by a set of Cartesian axes of coordinates. In this book, we shall always denote a stationary frame of reference by the symbol Σ, and a moving frame of reference by the symbol Σ'. A special case of a moving frame of reference is a frame of reference moving with constant speed along a straight line. Such a frame of reference is called the *inertial frame of reference*. In this book we shall only use inertial frames of reference.

Relativistic electromagnetism combines basic electromagnetic laws with the *principle of relativity*. The principle of relativity was first enunciated in 1638 by Galileo as a statement of the fact that there are no experiments or observations whereby one could distinguish the state of uniform motion along a straight line from the state of rest. However, in accordance with the level of scientific knowledge of his times, Galileo supported this statement by citing only mechanical experiments and observations with an indirect reference to the laws of optics. At the beginning of the 20th century, Lorentz, Poincaré, Larmor, and Einstein, in separate works, demonstrated that the principle of relativity was applicable to electromagnetic phenomena as well.

The expression *relativity theory* (or simply *relativity*), as it is now used in physics, has several different meanings. In particular, one differentiates between the *relativity theory of Lorentz and Poincaré*, *Einstein's special relativity theory*, and *Einstein's general relativity theory*.

Einstein's general relativity theory is his theory of gravitation and has little in common with other "relativities." The Lorentz-Poincaré relativity theory and Einstein's special relativity theory[1] have at least two things in common: they affirm the principle of relativity and they describe physical phenomena (mainly electromagnetic) associated with rapidly moving particles.

The significance of the Lorentz-Poincaré relativity theory, the significance of Einstein's special relativity theory, the difference between the two theories, and the allocation of priorities in the development of these theories are the subjects of considerable controversy.[2,3,4] However, there is no doubt that Einstein's special relativity theory is uniquely original insofar as the central point of the theory is the idea of the "relativity of space and time" closely associated with Einstein's concepts of "relativistic length contraction" and "relativistic time dilation."[5]

Relativistic electromagnetism and relativistic mechanics are usually presented in textbooks as consequences of Einstein's special relativity theory. However, in this book we shall use a novel approach to relativity, quite different from those used by Einstein, Lorentz, or Poincaré. We shall develop relativistic electromagnetism solely on the basis of electromagnetic retardation combined with the principle of relativity without any additional postulates, hypotheses, or conjectures. In turn, starting with relativistic electromagnetism, we shall develop relativistic mechanics, likewise without any additional postulates, hypotheses, or conjectures.

6-2. Equations for Transforming Electric and Magnetic Fields of Uniformly Moving Charge Distributions into Electric and Magnetic Fields of the Same Stationary Charge Distributions[6]

Consider a charge distribution of density ρ moving with constant velocity $\mathbf{v} = v_x\mathbf{i} = -v\mathbf{i}$. According to Eq. (5-1.11), the electric field of such a charge distribution is given by the present-time integral

$$\mathbf{E} = -\frac{1}{4\pi\varepsilon_0}\int\frac{\nabla\rho - \mathbf{i}(v^2/c^2)\partial\rho/\partial x'}{\{x'^2+(y'^2+z'^2)/\gamma^2\}^{1/2}}dV', \qquad (6\text{-}2.1)$$

or, factoring out γ,

$$\mathbf{E} = -\frac{\gamma}{4\pi\varepsilon_0}\int\frac{\nabla\rho - \mathbf{i}(v^2/c^2)\partial\rho/\partial x'}{(\gamma^2x'^2+y'^2+z'^2)^{1/2}}dV', \qquad (6\text{-}2.2)$$

where we use the standard abbreviation

$$\gamma = \frac{1}{(1 - v^2/c^2)^{1/2}}. \tag{6-2.3}$$

Now, since $\mathbf{J} = \rho\mathbf{v}$ and $\mathbf{v} = v_x\mathbf{i} = -v\mathbf{i}$, we have

$$-\mathbf{i}(v^2/c^2)\frac{\partial\rho}{\partial x'} = -(v/c^2)\frac{\partial(\rho v)}{\partial x'}\mathbf{i} = (v/c^2)\frac{\partial(\rho v_x)}{\partial x'}\mathbf{i}, \tag{6-2.4}$$

or

$$-\mathbf{i}(v^2/c^2)\frac{\partial\rho}{\partial x'} = (v/c^2)\frac{\partial J_x}{\partial x'}\mathbf{i}. \tag{6-2.5}$$

Equation (6-2.2) becomes therefore

$$\mathbf{E} = -\frac{\gamma}{4\pi\varepsilon_0}\int\frac{\nabla'\rho + (v/c^2)(\partial J_x/\partial x')\mathbf{i}}{(\gamma^2x'^2 + y'^2 + z'^2)^{1/2}}dV'. \tag{6-2.6}$$

The magnetic flux density field produced by the moving charge distribution is then, according to Eq. (3-2.6), taking into account that $\mathbf{v} \times \mathbf{i} = 0$ and using $1/c^2 = \varepsilon_0\mu_0$,

$$\begin{aligned}\mathbf{B} &= -\frac{\gamma\mu_0}{4\pi}\int\frac{\mathbf{v}\times\nabla'\rho}{(\gamma^2x'^2 + y'^2 + z'^2)^{1/2}}dV' \\ &= \frac{\gamma\mu_0}{4\pi}\int\frac{\nabla'\times\rho\mathbf{v}}{(\gamma^2x'^2 + y'^2 + z'^2)^{1/2}}dV',\end{aligned} \tag{6-2.7}$$

or, since $\rho\mathbf{v} = \rho v_x\mathbf{i} = J_x\mathbf{i}$,

$$\mathbf{B} = \frac{\gamma\mu_0}{4\pi}\int\frac{\nabla'\times J_x\mathbf{i}}{(\gamma^2x'^2 + y'^2 + z'^2)^{1/2}}dV'. \tag{6-2.8}$$

For the same *stationary* charge distribution, the field equations corresponding to Eqs. (6-2.6) and (6-2.8) are[7]

$$\mathbf{E} = -\frac{1}{4\pi\varepsilon_0}\int\frac{\nabla'\rho}{(x'^2+y'^2+z'^2)^{1/2}}dV' \qquad (6\text{-}2.9)$$

and

$$\mathbf{B} = 0. \qquad (6\text{-}2.10)$$

We shall now obtain a set of transformation equations which convert Eqs. (6-2.6) and (6-2.8) into Eqs. (6-2.9) and (6-2.10). The extraordinary significance of these transformation equations will become clear later, when we shall use them as the foundation for developing the theory of relativistic electromagnetism.

Since we are dealing with similar quantities relating to the moving and to the stationary charge distribution, we shall denote quantities pertaining to the moving charge distribution by subscript "m" and those pertaining to the stationary charge distribution by subscript "s," except when the relations are self-evident.

Let us write Eqs. (6-2.6) and (6-2.8) in terms of their Cartesian components. From Eq. (6-2.6) we have, resolving $\nabla'\rho$ into its Cartesian components,

$$E_{xm} = -\frac{\gamma}{4\pi\varepsilon_0}\int\frac{(\partial/\partial x')\{\rho+(v/c^2)J_x\}}{(\gamma^2x'^2+y'^2+z'^2)^{1/2}}dV', \qquad (6\text{-}2.11)$$

$$E_{ym} = -\frac{\gamma}{4\pi\varepsilon_0}\int\frac{\partial\rho/\partial y'}{(\gamma^2x'^2+y'^2+z'^2)^{1/2}}dV', \qquad (6\text{-}2.12)$$

$$E_{zm} = -\frac{\gamma}{4\pi\varepsilon_0}\int\frac{\partial\rho/\partial z'}{(\gamma^2x'^2+y'^2+z'^2)^{1/2}}dV'. \qquad (6\text{-}2.13)$$

From Eq. (6-2.8) we similarly have, resolving $\nabla'\times J_x\mathbf{i}$ into its Cartesian components,

$$B_{xm} = 0, \qquad (6\text{-}2.14)$$

$$B_{ym} = \frac{\gamma\mu_0}{4\pi}\int \frac{\partial J_x/\partial z'}{(\gamma^2 x'^2 + y'^2 + z'^2)^{1/2}} dV', \qquad (6\text{-}2.15)$$

$$B_{zm} = -\frac{\gamma\mu_0}{4\pi}\int \frac{\partial J_x/\partial y'}{(\gamma^2 x'^2 + y'^2 + z'^2)^{1/2}} dV'. \qquad (6\text{-}2.16)$$

Let us also write Eqs. (6-2.9) and (6-2.10), representing the electric and magnetic fields of the stationary charge distribution, in terms of the Cartesian components. From Eq. (6-2.9) we have

$$E_{xs} = -\frac{1}{4\pi\varepsilon_0}\int \frac{\partial\rho/\partial x'}{(x'^2 + y'^2 + z'^2)^{1/2}} dV', \qquad (6\text{-}2.17)$$

$$E_{ys} = -\frac{1}{4\pi\varepsilon_0}\int \frac{\partial\rho/\partial y'}{(x'^2 + y'^2 + z'^2)^{1/2}} dV', \qquad (6\text{-}2.18)$$

$$E_{zs} = -\frac{1}{4\pi\varepsilon_0}\int \frac{\partial\rho/\partial z'}{(x'^2 + y'^2 + z'^2)^{1/2}} dV'. \qquad (6\text{-}2.19)$$

From Eq. (6-2.10) we have

$$B_{xs} = B_{ys} = B_{zs} = 0. \qquad (6\text{-}2.20)$$

The transformations that we seek are those that transform Eqs. (6-2.11)-(6-2.16) into Eqs. (6-2.17)-(6-2.20).

Clearly, to achieve the desired transformations, we need to transform the denominators of the integrands in Eqs. (6-2.11)-(6-2.13) into the denominators of the integrands in Eqs. (6-2.17)-(6-2.19). Comparing Eqs. (6-2.11)-(6-2.13) with Eqs. (6-2.17)-(6-2.19), we recognize that the desired transformation of the denominators will be achieved if we set

$$x'_s = \gamma x'_m, \qquad (6\text{-}2.21)$$

$$y'_s = y'_m, \tag{6-2.22}$$

$$z'_s = z'_m, \tag{6-2.23}$$

because the denominators of the integrands in Eqs. (6-2.11)-(6-2.16) can then be written as $(x_s'^2 + y_s'^2 + z_s'^2)^{1/2} = (x'^2 + y'^2 + z'^2)_s^{1/2}$ and thus become the same as in Eqs. (6-2.17)-(6-2.19). Observe that this transformation simply changes the scale units of the x axis for the stationary charge distribution and does not alter the physical significance of Eqs. (6-2.17)-(6-2.20). However, if we change the scale units of the x axis for the stationary charge distribution, then the derivatives $\partial/\partial x'$ in Eqs. (6-2.11) and (6-2.17) are no longer equal. According to Eq. (6-2.21), the correlation between them is now

$$\left(\frac{\partial}{\partial x'}\right)_s = \frac{1}{\gamma}\left(\frac{\partial}{\partial x'}\right)_m. \tag{6-2.24}$$

Likewise, the volume elements $dV' = dx'dy'dz'$ in the equations for the moving and for the stationary charge distribution are no longer the same. The correlation between them is now

$$dV_s' = \gamma dV_m'. \tag{6-2.25}$$

If we now substitute Eqs. (6-2.21)-(6-2.25) into Eqs. (6-2.11)-(6-2.16), we obtain, using subscripts "s" and "m" in the integrands to keep track of the transformation steps,

$$E_{xm} = -\frac{\gamma}{4\pi\varepsilon_0}\int \frac{(\partial/\partial x')_s\{\rho + (v/c^2)J_x\}_m}{(x'^2 + y'^2 + z'^2)_s^{1/2}}\, dV_s', \tag{6-2.26}$$

$$E_{ym} = -\frac{1}{4\pi\varepsilon_0}\int \frac{(\partial/\partial y')_s\rho_m}{(x'^2 + y'^2 + z'^2)_s^{1/2}}\, dV_s', \tag{6-2.27}$$

$$E_{zm} = -\frac{1}{4\pi\varepsilon_0}\int \frac{(\partial/\partial z')_s \rho_m}{(x'^2+y'^2+z'^2)_s^{1/2}} dV_s' \, ; \qquad (6\text{-}2.28)$$

$$B_{xm} = 0, \qquad (6\text{-}2.29)$$

$$B_{ym} = \frac{\mu_0}{4\pi}\int \frac{(\partial/\partial z')_s J_{xm}}{(x'^2+y'^2+z'^2)_s^{1/2}} dV_s' \, , \qquad (6\text{-}2.30)$$

$$B_{zm} = -\frac{\mu_0}{4\pi}\int \frac{(\partial/\partial y')_s J_{xm}}{(x'^2+y'^2+z'^2)_s^{1/2}} dV_s' \, . \qquad (6\text{-}2.31)$$

Comparing the numerators in Eq. (6-2.17) and Eq. (6-2.26), we immediately recognize that the equation for transforming the numerator in Eq. (6-2.26) into the numerator in Eq. (6-2.17) is

$$\rho_s = \gamma\{\rho + (v/c^2)J_x\}_m . \qquad (6\text{-}2.32)$$

Substituting Eq. (6-2.32) into Eq. (6-2.26), we obtain

$$E_{xm} = -\frac{1}{4\pi\varepsilon_0}\int \frac{(\partial/\partial x')_s \rho_s}{(x'^2+y'^2+z'^2)_s^{1/2}} dV_s' \, . \qquad (6\text{-}2.33)$$

All we now need to complete the transformation of Eq. (6-2.11) into Eq. (6-2.17) is to replace E_{xm} on the left of Eq. (6-2.33) by E_{xs}. We denote this transformation step by the field transformation equation

$$E_{xs} = E_{xm} . \qquad (6\text{-}2.34)$$

Examining the remaining Eqs. (6-2.27)-(6-2.31), we recognize that in order to use Eq. (6-2.32) with these equations we need to combine equations for E_m and B_m. Noting that $\mu_0\varepsilon_0 = 1/c^2$, combining Eqs. (6-2.27)-(6-2.28) with Eqs. (6-2.30)-(6-2.31) so that the expression $\gamma\{\rho + (v/c^2)J_x\}$ appears in the combined equations, using Eqs. (6-2.18) and (6-2.19) as the transformation "targets," and using Eq. (6-2.32), we recognize that the remaining transformation equations for the electric field must be

$$E_{ys} = \gamma(E_y + vB_z)_m, \tag{6-2.35}$$

$$E_{zs} = \gamma(E_z - vB_y)_m. \tag{6-2.36}$$

Examining again Eqs. (6-2.27)-(6-2.31), remembering that $J_x = -v\rho$, and using Eq. (6-2.20) as the transformation "target", we tentatively identify the transformation equations for the magnetic field as

$$B_{xs} = B_{xm}, \tag{6-2.37}$$

$$B_{ys} = \gamma(B_y - vE_z/c^2)_m, \tag{6-2.38}$$

$$B_{zs} = \gamma(B_z + vE_y/c^2)_m \tag{6-2.39}$$

(these equations must be considered tentative because the factor γ in them is as yet uncertain; the need for it will be established in the next section).

As was stated in Sections 4-1, 4-5, and 5-3 (see Examples 4-1.2, 4-5.1, and 5-3.1), Eqs. (6-2.6), (6-2.7), and the subsequent equations for the fields of the moving charge distribution are "snapshots" representing instantaneous fields of the charge distribution observed at $t = 0$. Therefore also Eq. (6-2.21) is only valid for $t = 0$, so that $x_s' = \gamma x_{m,t=0}$. We shall now put Eq. (6-2.21) into a more general form by assuming that the time of observation is an unspecified t. Since the charge distribution moves with velocity v in the negative direction of the x axis, the present position of the distribution is shifted toward smaller values of x, in accordance with

$$x_m' = (x_{t=0}' - vt)_m \tag{6-2.40}$$

as t increases. Hence, for a general case, Eq. (6-2.21) must be replaced by

$$x_s' = \gamma(x' + vt)_m, \tag{6-2.41}$$

where x' is the x coordinate of dV'_m at the time t. Note that although x_s' in Eq. (6-2.41) appears to depend on time, in reality it does not depend on time since by Eq. (6-2.40) $(x' + vt)_m = x'_{m,t=0}$.

6-3. Inverse Transformations

According to the principle of relativity, it is impossible to tell whether the charge which we call "moving" really moves with velocity $\mathbf{v} = -v\mathbf{i}$ relative to our laboratory and relative to the charge that we call "stationary," or whether the laboratory with the charge which we call "stationary" moves with velocity $\mathbf{v} = v\mathbf{i}$ relative to the charge that we call "moving." Consequently, the transformation equations obtained in Section 6-2 should be applicable not only for transforming the fields of a moving charge distribution into the fields of a stationary charge distribution, but also for transforming the fields of a stationary charge distribution into the fields of a moving charge distribution by simply reversing the sign in front of v and transposing the subscripts m and s. From Eqs. (6-2.40), (6-2.22), (6-2.23), (6-2.32), and (6-2.34)-(6-2.39) we obtain therefore the following set of inverse transformation equations (equations for transforming fields of a stationary charge distribution into the fields of the same moving charge distribution)

$$x'_m = \gamma(x' - vt)_s, \tag{6-3.1}$$

$$y_m' = y_s', \tag{6-3.2}$$

$$z_m' = z_s', \tag{6-3.3}$$

$$\rho_m = \gamma\{\rho - (v/c^2)J_x\}_s, \tag{6-3.4}$$

$$E_{xm} = E_{xs}, \tag{6-3.5}$$

$$E_{ym} = \gamma(E_y - vB_z)_s, \tag{6-3.6}$$

$$E_{zm} = \gamma(E_z + vB_y)_s, \tag{6-3.7}$$

$$B_{xm} = B_{xs}, \tag{6-3.8}$$

$$B_{ym} = \gamma(B_y + vE_z/c^2)_s, \tag{6-3.9}$$

$$B_{zm} = \gamma(B_z - vE_y/c^2)_s. \tag{6-3.10}$$

Observe that Eqs. (6-3.6), (6-3.7), (6-3.9), and (6-3.10) can also be obtained by solving Eq. (6-2.35), (6-2.36), (6-2.38), and (6-2.39) for the components of E_m and B_m in terms of the components of E_s and B_s. Eliminating B_{zm} between Eqs. (6-2.35) and (6-2.39), eliminating B_{ym} between Eqs. (6-2.36) and (6-2.38)), eliminating E_{zm} between Eqs. (6-2.36) and (6-2.38), and eliminating E_{ym} between Eqs. (6-2.35) and (6-2.39), we obtain Eqs. (6-3.6), (6-3.7), (6-3.9), and (6-3.10) directly, without invoking the principle of relativity. However, the equations so obtained clearly confirm the principle of relativity for electric and magnetic fields.

We shall now supplement our transformation equations by four more equations. Solving Eqs. (6-2.32) and (6-3.4) for J_x, we obtain

$$J_{xs} = \gamma(J_x + v\rho)_m \tag{6-3.11}$$

and

$$J_{xm} = \gamma(J_x - v\rho)_s. \tag{6-3.12}$$

Solving Eqs. (6-2.40) and (6-3.1) for t_s and t_m, we obtain

$$t_s = \gamma(t + vx/c^2)_m, \tag{6-3.13}$$

and

$$t_m = \gamma(t - vx'/c^2)_s. \tag{6-3.14}$$

Let us now prove the need for the factor γ in Eqs. (6-2.38) and (6-2.39). According to Eq. (3-2.6), a uniformly moving charge distribution always creates a magnetic flux density field given by

$$\mathbf{B}_m = (\mathbf{v} \times \mathbf{E}_m)/c^2, \tag{6-3.15}$$

where $\mathbf{E}_m$ is the electric field produced by the moving charge distribution. Consider, for example, the y component of $\mathbf{B}_m$. If $\mathbf{v} = -v\mathbf{i}$, then by Eq. (6-3.15), this component is

$$B_{ym} = vE_{zm}/c^2. \tag{6-3.16}$$

But by Eq. (6-3.9), the same component is (noting that for a stationary charge $B_{ys} = 0$)

$$B_{ym} = \gamma vE_{zs}/c^2. \tag{6-3.17}$$

Now, according to Eq. (6-3.7), $\gamma E_{zs} = E_{zm}$, so that Eq. (6-3.17) can be written as

$$B_{ym} = vE_{zm}/c^2, \tag{6-3.18}$$

which is exactly the same as Eq. (6-3.16). Clearly, if the factor γ were not present in Eq. (6-3.9), then the factor $1/\gamma$ would appear in Eq. (6-3.18), the agreement between Eq. (6-3.16) and Eq. (6-3.18) would not be possible, and therefore our transformation equations for $\mathbf{B}$ would be incorrect. But since Eq. (6-3.9) has been obtained from Eq. (6-2.38), the factor γ must be present also in Eq. (6-2.38) [and therefore in Eq. (6-2.39) as well].

Note that none of the transformation equations obtained in this and in the preceding section of the book constitute actual equalities between the quantities involved. These equations are merely prescriptions for obtaining electric and magnetic fields of a stationary charge distribution from the fields of the same moving charge distribution by replacing quantities pertaining to the moving charge distribution by quantities pertaining to the stationary charge distribution and vice versa.

6-4. Equations for Transforming Electric and Magnetic Potentials of Uniformly Moving Charge Distributions into Electric and Magnetic Potentials of the Same Stationary Charge Distributions and Vice Versa

According to Eq. (5-2.4), the electric scalar potential of a uniformly moving charge distribution is given by the present-time integral

$$\varphi = \frac{1}{4\pi\varepsilon_0}\int \frac{\rho}{\{x'^2 + (y'^2 + z'^2)/\gamma^2\}^{1/2}}\, dV', \tag{6-4.1}$$

or, factoring out γ,

$$\varphi = \frac{\gamma}{4\pi\varepsilon_0}\int \frac{\rho}{(\gamma^2 x'^2 + y'^2 + z'^2)^{1/2}}\, dV' \tag{6.4.2}$$

and, according to Eq. (5-2.16), the magnetic vector potential is given by the present-time integral

$$\mathbf{A} = \frac{\mathbf{v}}{4\pi\varepsilon_0 c^2}\int \frac{\rho}{\{x'^2 + (y'^2 + z'^2)/\gamma^2\}^{1/2}}\, dV', \tag{6-4.3}$$

or

$$\mathbf{A} = \frac{\gamma\mathbf{v}}{4\pi\varepsilon_0 c^2}\int \frac{\rho}{(\gamma^2 x'^2 + y'^2 + z'^2)^{1/2}}\, dV'. \tag{6-4.4}$$

For the same stationary charge distribution, the potential equations corresponding to Eqs. (6-4.2) and (6-4.4) are[8]

$$\varphi = \frac{1}{4\pi\varepsilon_0}\int \frac{\rho}{(x'^2 + y'^2 + z'^2)^{1/2}}\, dV' \tag{6-4.5}$$

and

$$\mathbf{A} = 0. \tag{6-4.6}$$

We shall now obtain a set of simple transformation equations that convert Eqs. (6-4.2) and (6-4.4) into Eqs. (6-4.5) and (6-4.6).

As before, since we are dealing with similar quantities relating to the moving and to the stationary charge distribution, we shall denote quantities pertaining to the moving charge distribution by subscript "m" and those pertaining to the stationary charge distribution by subscript "s," except when the relations are self-evident.

Let us write Eqs (6-4.4) and (6-4.6) in terms of their Cartesian components. Assuming, as before, that the charge distribution under consideration moves with velocity $\mathbf{v} = -v\mathbf{i}$, and using $v_x\rho = J_x$, we have from Eq. (6-4.4)

$$A_{xm} = \frac{\gamma}{4\pi\varepsilon_0 c^2}\int \frac{J_x}{(\gamma^2 x'^2 + y'^2 + z'^2)^{1/2}}\, dV', \tag{6-4.7}$$

$$A_{ym} = 0, \tag{6-4.8}$$

$$A_{zm} = 0. \tag{6-4.9}$$

From Eq. (6-4.6), we obtain

$$A_{xs} = A_{ys} = A_{zs} = 0. \tag{6-4.10}$$

We seek transformation equations that convert Eq. (6-4.2) into Eq. (6-4.5) and Eqs. (6-4.7)-(6-4.9) into Eq. (6-4.10). Clearly, to achieve the desired transformations, we need to transform the denominator in the integrand of Eq. (6-4.2) into the denominator of Eqs. (6-4.5). However, we have already found that this transformation can be achieved by using Eqs. (6-2.21)-(6-2.23). Of course, if we use Eqs. (6-2.21)-(6-2.23), then we must also use Eq. (6-2.25) for transforming the volume elements in the integrals of Eqs. (6-4.2), (6-4.5), and (6-4.7). Naturally, we want to use as few transformation equations for all electric and magnetic quantities and formulas as possible, and since we have already obtained Eq. (6-2.32) for transforming ρ, we shall use Eq. (6-2.32) now again.

Using Eqs. (6-2.21)-(6-2.23), (6-2.25), and (6-2.32) for substituting x', y', z', dV', and ρ in Eq. (6-4.5) , we obtain

$$\varphi_s = \frac{\gamma^2}{4\pi\varepsilon_0}\int \frac{\{\rho + (v/c^2)J_x\}_m}{(\gamma^2 x'^2 + y'^2 + z'^2)_m^{1/2}}\, dV'_m. \tag{6-4.11}$$

Examining Eq. (6-4.11), we recognize that the integral in it is a combination of Eqs. (6-4.2) and (6-4.7), so that Eq. (6-4.11) can be written as

$$\varphi_s = \gamma(\varphi + vA_x)_m, \tag{6-4.12}$$

which is the desired transformation equation for the scalar potential.

Remembering that $v\rho = -J_x$ and combining Eqs. (6-4.2) and (6-4.7) so that $A_{xs} = 0$ [as required by Eq. (6-4.10)], we find that the transformation equation for the x component of the magnetic vector potential is

$$A_{xs} = \gamma\{A_x + (v/c^2)\varphi\}_m \tag{6-4.13}$$

(this equation must be considered tentative because the factor γ in it is as yet uncertain; we shall prove the need for it shortly).

For the remaining components of **A**, we obtain by comparing Eqs. (6-4.8), (6-4.9), and (6-4.10)

$$A_{ys} = A_{ym}. \tag{6-4.14}$$

$$A_{zs} = A_{zm}. \tag{6-4.15}$$

As in the case of transformation equations for electric and magnetic fields, the relativity principle demands that the inverse transformation equations should be obtainable from Eqs. (6-4.12)-(6-4.15) by simply reversing the sign in front of v and transposing the subscripts "s" and "m." The inverse transformation equations are therefore

$$\varphi_m = \gamma(\varphi - vA_x)_s, \tag{6-4.16}$$

$$A_{xm} = \gamma[A_x - (v/c^2)\varphi]_s, \tag{6-4.17}$$

$$A_{ym} = A_{ys}. \tag{6-4.18}$$

$$A_{zm} = A_{zs}. \tag{6-4.19}$$

Let us now prove the need for the factor γ in Eqs. (6-4.13) and (6-4.17). According to Eq. (5-2.12), the magnetic and electric potentials of a uniformly moving charge distribution are connected by the equation

$$\mathbf{A}_m = \mathbf{v}\varphi_m/c^2. \tag{6-4.20}$$

Consider the x component of $\mathbf{A}_m$. If $\mathbf{v} = -v\mathbf{i}$, then by Eq. (6-4.20), this component is

$$A_{xm} = -v\varphi_m/c^2. \tag{6-4.21}$$

But by Eq. (6-4.17), the same component is (noting that for a stationary charge $A_{xs} = 0$)

$$A_{xm} = -\gamma v\varphi_s/c^2. \tag{6-4.22}$$

Now, according to Eq. (6-4.16), $\gamma\varphi_s = \varphi_m$, so that Eq. (6-4.22) can be written as

$$A_{xm} = -v\varphi_m/c^2, \tag{6-4.23}$$

which is exactly the same as Eq. (6-4.21). Clearly, if the factor γ were not present in Eq. (6-4.17), then the factor $1/\gamma$ would appear in Eq. (6-4.23), the agreement between Eq. (6-4.21) and Eq. (6-4.23) would not be possible, and therefore our transformation equations for $\mathbf{A}$ would be incorrect. But since Eq. (6-4.17) has been obtained from Eq. (6-4.13), the factor γ must be present also in Eq. (6-4.13).

Note that none of the transformation equations obtained in this section of the book constitute actual equalities between the quantities involved. These equations are merely prescriptions for obtaining electric and magnetic potentials of a stationary charge distribution from the potentials of the same moving charge distribution by replacing quantities pertaining to the moving charge distribution by quantities pertaining to the stationary charge distribution and vice versa.

References and Remarks for Chapter 6

1. The name "relativity theory" ("Relativtheorie" in German) was coined by Max Planck as an abbreviation for the Lorentz-Einstein ("Lorentz-Einsteinsche") electrodynamic theory and its application to the motion of the electron [see Max Planck, "Die Kaufmannschen Messungen der Ablenkbarkeit der β-Strahlen in ihrer Bedeutung für die Dynamik der Elektronen," Phys. Z. **7**, 753-761 (1906); A. H. Bucherer, in the discussion section of this article, called Einstein's theory the "Relativitätstheorie"]. Einstein used the name "relativity theory" ("Relativitätstheorie") for the first time in his article "Die vom Relativitätsprinzip geforderte Trägheit der Energie," Ann. Phys. **23**, 371-384 (1907).

2. According to E. T. Whittaker (author of the highly respected *A History of the Theories of Aether and Electricity*), Einstein's contribution to relativity theory was minimal. Referring to Einstein's famous article "Zur Elektrodynamik bewegter Körper," Ann. Phys. **17**, 891-921 (1905), Whittaker says : "In the autumn of the same year [1905]. . . , Einstein published a paper which set forth the relativity theory of Poincaré and Lorentz with some amplifications, and which attracted much attention" [see E. T. Whittaker, *A History of the Theories of Aether and Electricity* (Thomas Nelson, London, 1953) Vol. II, Chapt. 2 ("The Relativity Theory of Poincaré and Lorentz") p 40]. Whittaker's assessment is contrasted, for example, with that by Arthur I. Miller [author of the very detailed "biography and analysis of the (Einstein's) relativity paper set into its historical context"]. Miller describes Einstein's 1905 article as follows: "Page for page Einstein's relativity paper is unparalleled in the history of science in its depth, breadth and sheer intellectual virtuosity. . . the consequences of the special relativity theory changed mankind's very view of its relation to cosmos. . ." [see Arthur I. Miller, "*Albert Einstein's Special Theory of Relativity*, (Addison-Wesley, Reading, Massachusetts, 1981) p. xiii].

3. According to K. N. Schaffner (a very thorough investigator of the history of Lorentz's and Einstein's relativity theories) ". . . it is clear that Lorentz's theory and Einstein's theory are rather different theories — but it is exceedingly difficult precisely to define the difference" [Kenneth F. Schaffner, "The Lorentz Electron Theory of Relativity," Am. J. Phys. **37**, 498-513 (1969)]. See also Charles Scribner, Jr., "Henri Poincaré and the Principle of Relativity," Am. J. Phys. **32**, 672-678 (1964); Stanley Goldberg, "Henri Poincaré and Einstein's Theory of Relativity," Am. J. Phys. **35**, 934-944 (1967); C. Kittel, "Larmor and the Prehistory of the Lorentz Transformation," Am. J. Phys. **42**, 726-729 (1974).

4. The controversy is partly caused by the fact that neither of the two articles on relativity published by Einstein in 1905 [the first article was "Zur Elektrodynamik bewegter Körper," Ann. Phys. **17**, 891-921 (1905), the second article was "Ist die Trägheit eines Körpers von seinem Energieinhalt abhängig?," Ann. Phys. **18**, 639-641 (1905)] has any references to works by other authors, although Lorentz transformations of coordinates and time, transformations of electric and magnetic fields, etc., which Einstein used in his first paper were well known in 1905 from the works of Lorentz, Larmor, and Poincaré (see Refs. 1 and 2 in Chapter 7). In this connection it is noteworthy that the editors of the *Collected Papers* of H. Poincaré specifically pointed out that the method of clock synchronization by means of light signals used by Einstein in his first relativity article was due to Poincaré. They also stated that from the mathematical point of view Einstein's 1905 article presented nothing more than what had been published by Lorentz and Poincaré ("Le célèbre Mémoir de A. Einstein *Zur Elektrodynamik bewegter Körper* n'apportant rien de plus au point de vue mathématique que les publications de H. A. Lorentz et de H. Poincaré") [see *Ouvres de Henri Poincaré*, (Gauthier-Villars, Paris, 1954) Vol. IX, pp. 698, 699 (this volume contains most of Poincaré's papers pertaining to relativity); see also *Handbuch der Physik*, (Springer, Berlin, 1927) Vol. XII, p. 270]. It is also noteworthy that A. I. Miller in his comprehensive book on the history of Einstein's special relativity theory (see Ref. 2, above) decided not to discuss Larmor's contribution to relativity theory "because in my opinion Larmor's work had an indirect effect, if any, on Lorentz's thinking toward the electron theory of 1904" (p. 114). Taking into account that as early as 1900 Larmor, in his book *Aether and Matter* (Cambridge U. P., Cambridge, 1900), published (in his own notation) all basic relativistic transformation equations for time and space coordinates and for electromagnetic quantities which Einstein presented in his first 1905 article, and that in 1895 Poincaré devoted a large article (in four separate parts) to Larmor's earlier work, Miller's decision only perpetuates the controversy.

5. See Albert Einstein, *The Meaning of Relativity*, (Princeton University Press, Princeton, New Jersey, 1950), p. 30, 31, 36 and A. Einstein "Die Relativitätstheorie" in E. Lecher, ed., *Physik*, 2nd ed., (Teubner, Leipzig, 1925) p. 791-793.

6. See also Oleg D. Jefimenko, "Retardation and relativity: Derivation of Lorentz-Einstein transformations from retarded integrals for electric and magnetic fields," Am. J. Phys. **63**, 267-272 (1995).

7. Oleg D. Jefimenko, *Electricity and Magnetism*, 2nd ed., (Electret Scientific, Star City, 1989) pp. 93 and 343. Observe that Eqs. (6-2.9) and (6-2.10) can be obtained from Eqs. (6-2.6) and (6-2.8) by setting $\mathbf{v} = 0$ and $\mathbf{J} = 0$.

8. See, for example, Ref. 7, pp. 120 and 364. Observe that Eqs. (6-4.5) and (6-4.6) can be obtained from Eqs. (6-4.2) and (6-4.4) by setting $\mathbf{v} = 0$.

7

THE ESSENTIALS OF RELATIVISTIC ELECTRODYNAMICS

Relativistic electrodynamics provides powerful yet simple methods for solving a variety of problems involving uniformly moving electromagnetic systems. In this chapter we shall familiarize ourselves with the basic equations of relativistic electrodynamics, their properties, consequences, and methods of their application.

7-1. Basic Relativistic Transformation Equations

The basic equations of relativistic electrodynamics are the transformation equations for coordinates, time, and electromagnetic quantities derived in Chapter 6. However, in relativistic electrodynamics these equations have a somewhat different physical meaning and are customarily expressed in a notation different from the notation used in Chapter 6.

To convert the equations derived in Chapter 6 into the standard relativistic form, we shall now assume that the stationary charge distribution used in Chapter 6 is located in a reference frame Σ' uniformly moving with respect to the laboratory (reference frame Σ). Since the charge is at rest in Σ', all quantities with subscript "s" used in Chapter 6 apply now to measurements performed in that reference frame. And since the reference frame Σ' together with the charge distribution moves with respect to the laboratory (reference frame Σ), all quantities with subscripts "m" apply now to measurements performed in the laboratory. The transformation equations derived in Chapter 6 applied to a charge distribution moving with a velocity $\mathbf{v} = -v\mathbf{i}$, that is, in the negative direction of the x axis. In relativistic electrodynamics the reference frame Σ' is usually assumed to move with a velocity $\mathbf{v} = v\mathbf{i}$, that is, in the positive direction of the x axis, and both frames Σ and Σ' are assumed to have a common x axis and a common xy plane (Fig. 7.1). Furthermore, in relativistic electrodynamics the quantities pertaining to the moving and the stationary charge distribution are customarily designated not by means of subscripts, but by using primes for identifying the quantities measured in the moving frame Σ' and by using ordinary notation for the quantities measured in the laboratory.

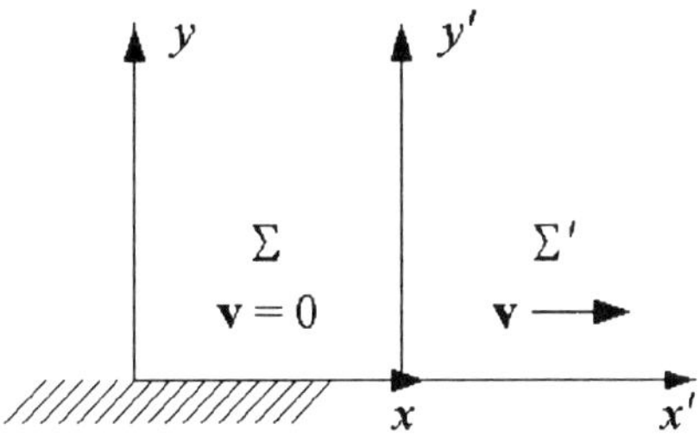

Fig. 7.1 Reference frame Σ' moves with velocity $\mathbf{v}$ with respect to the laboratory (reference frame Σ).

To put the transformation equations obtained in Chapter 6 into the customary relativistic form, we need therefore to modify these equations as follows: omit the subscript "m," replace the subscript "s" by a prime, and reverse the sign in front of v. Observe that we no longer can denote the field point coordinates by primes, since the primes must now be used only for denoting quantities measured in the moving reference system. Therefore, before making any other modifications, we must first remove the primes from the transformation equations obtained in Chapter 6. Finally, since we shall not use retarded quantities in relativistic equations, we shall use the square brackets in relativistic equations and elsewhere as the ordinary algebraic symbols.

After making the indicated changes of notation, we then obtain for the quantities measured in Σ expressed in terms of the quantities measured in Σ':

(a) For the space and time coordinates

$$x = \gamma(x' + vt'), \tag{7-1.1}$$

$$y = y', \tag{7-1.2}$$

$$z = z', \tag{7-1.3}$$

$$t = \gamma(t' + vx'/c^2). \tag{7-1.4}$$

(b) For the electric field

$$E_x = E_x', \tag{7-1.5}$$

$$E_y = \gamma(E_y' + vB_z'), \tag{7-1.6}$$

$$E_z = \gamma(E_z' - vB_y'). \tag{7-1.7}$$

(c) For the magnetic flux density field

$$B_x = B_x', \tag{7-1.8}$$

$$B_y = \gamma(B_y' - vE_z'/c^2), \tag{7-1.9}$$

$$B_z = \gamma(B_z' + vE_y'/c^2). \tag{7-1.10}$$

(d) For the charge and current densities

$$\rho = \gamma[\rho' + (v/c^2)J_x'], \tag{7-1.11}$$

$$J_x = \gamma(J_x' + v\rho'), \tag{7-1.12}$$

$$J_y = J_y', \tag{7-1.13}$$

$$J_z = J_z'. \tag{7-1.14}$$

[Eqs. (7-1.13) and (7-1.14) follow from the fact that J_y and J_z do not enter into the transformation equations obtained in Chapter 6].

(e) For the scalar and vector potentials

$$\varphi = \gamma(\varphi' + vA_x'), \tag{7-1.15}$$

$$A_x = \gamma[A_x' + (v/c^2)\varphi'], \tag{7-1.16}$$

$$A_y = A_y', \tag{7-1.17}$$

$$A_z = A_z'. \tag{7-1.18}$$

For the quantities measured in Σ' expressed in terms of the quantities measured in Σ we similarly obtain:

(a) For the space and time coordinates

$$x' = \gamma(x - vt), \tag{7-1.19}$$

$$y' = y, \tag{7-1.20}$$

$$z' = z, \tag{7-1.21}$$

$$t' = \gamma(t - vx/c^2). \tag{7-1.22}$$

(b) For the electric field

$$E_x' = E_x, \tag{7-1.23}$$

$$E_y' = \gamma(E_y - vB_z), \tag{7-1.24}$$

$$E_z' = \gamma(E_z + vB_y). \tag{7-1.25}$$

(c) For the magnetic flux density field

$$B_x' = B_x, \tag{7-1.26}$$

$$B_y' = \gamma(B_y + vE_z/c^2), \tag{7-1.27}$$

$$B_z' = \gamma(B_z - vE_y/c^2). \tag{7-1.28}$$

(d) For charge and current densities

$$\rho' = \gamma[\rho - (v/c^2)J_x], \tag{7-1.29}$$

$$J'_x = \gamma(J_x - v\rho), \tag{7-1.30}$$

$$J'_y = J_y, \tag{7-1.31}$$

$$J'_z = J_z. \tag{7-1.32}$$

(e) For the scalar and vector potentials

$$\varphi' = \gamma(\varphi - vA_x), \tag{7-1.33}$$

$$A_x' = \gamma[A_x - (v/c^2)\varphi], \tag{7-1.34}$$

$$A_y' = A_y, \tag{7-1.35}$$

$$A_z' = A_z. \tag{7-1.36}$$

The relativistic transformation equations for coordinates, time, and electric and magnetic fields are usually called the *Lorentz transformation equations.*[1] The relativistic transformation equations for electric and magnetic fields together with the transformation equations for the electric charge and current density are sometimes called the *Lorentz-Einstein transformation equations.*[2] The relativistic transformation equations for scalar and vector potentials are due to Poincaré[3] but do not carry his name.

In the derivations that follow, we shall frequently use "hybrid" transformation equations obtained from the "regular" transformation equations listed above by transposing their terms so that an unprimed or a primed quantity becomes associated with both a primed quantity and an unprimed quantity. An example of such a hybrid equation is

$$E_y = E_y'/\gamma + vB_z \tag{7-1.37}$$

obtained from Eq. (7-1.24).

7-2. Transformation Equations for Velocity and Acceleration

Since relativistic electrodynamics is primarily concerned with moving electromagnetic systems, we need to know how to transform velocity and acceleration from one reference frame to another.

Let us first obtain transformation equations for velocity.[4] Let an object move with a velocity whose x, y, and z components measured in the rest frame Σ are $dx/dt = u_x$, $dy/dt = u_y$, and $dz/dt = u_z$. Let the corresponding components measured in the moving frame Σ' be $dx'/dt' = u_x'$, $dy'/dt' = u_y'$, and $dz'/dt' = u_z'$. Differentiating Eqs. (7-1.1)-(7-1.4), we have

$$dx = \gamma(dx' + v dt') = \gamma(u_x' + v)dt' , \tag{7-2.1}$$

$$dy = dy' , \tag{7-2.2}$$

$$dz = dz' , \tag{7-2.3}$$

$$dt = \gamma(dt' + v dx'/c^2) = \gamma(1 + v u_x'/c^2)dt' . \tag{7-2.4}$$

Dividing Eqs. (7-2.1)-(7-2.3) by Eq. (7-2.4), we obtain transformation equations for the velocity

$$u_x = \frac{u_x' + v}{1 + v u_x'/c^2} , \tag{7-2.5}$$

$$u_y = \frac{u_y'}{\gamma(1 + v u_x'/c^2)} , \tag{7-2.6}$$

$$u_z = \frac{u_z'}{\gamma(1 + v u_x'/c^2)} . \tag{7-2.7}$$

The inverse transformation equations are obtained, as usual, by transposing the primes and changing the sign in front of v.

They are

$$u_x' = \frac{u_x - v}{1 - vu_x/c^2}, \tag{7-2.8}$$

$$u_y' = \frac{u_y}{\gamma(1 - vu_x/c^2)}, \tag{7-2.9}$$

$$u_z' = \frac{u_z}{\gamma(1 - vu_x/c^2)}. \tag{7-2.10}$$

Let us now obtain transformation equations for an acceleration.[5] Let an object move with an acceleration whose x, y, and z components measured in the rest frame Σ are $du_x/dt = a_x$, $du_y/dt = a_y$, and $du_z/dt = a_z$. Let the corresponding components measured in the moving frame Σ' be $du_x'/dt' = a_x'$, $du_y'/dt' = a_y'$, and $du_z'/dt' = a_z'$.

Differentiating Eqs. (7-2.5)-(7-2.7), we obtain

$$du_x = \frac{(1 + vu_x'/c^2)du_x' - (u_x' + v)vdu_x'/c^2}{(1 + vu_x'/c^2)^2} = \frac{(1 - v^2/c^2)du_x'}{(1 + vu_x'/c^2)^2}$$
$$= \frac{du_x'}{\gamma^2(1 + vu_x'/c^2)^2}, \tag{7-2.11}$$

$$du_y = \frac{(1 + vu_x'/c^2)du_y' - u_y'vdu_x'/c^2}{\gamma(1 + vu_x'/c^2)^2} = \frac{du_y'}{\gamma(1 + vu_x'/c^2)} - \frac{u_y'vdu_x'/c^2}{\gamma(1 + vu_x'/c^2)^2}, \tag{7-2.12}$$

$$du_z = \frac{(1 + vu_x'/c^2)du_z' - u_z'vdu_x'/c^2}{\gamma(1 + vu_x'/c^2)^2} = \frac{du_z'}{\gamma(1 + vu_x'/c^2)} - \frac{u_z'vdu_x'/c^2}{\gamma(1 + vu_x'/c^2)^2}. \tag{7-2.13}$$

Dividing Eqs. (7-2.11)-(7-2.13) by Eq. (7-2.4), we obtain transformation equations for the acceleration

$$a_x = \frac{a_x'}{\gamma^3(1 + vu_x'/c^2)^3}, \tag{7-2.14}$$

$$a_y = \frac{a_y'}{\gamma^2(1 + vu_x'/c^2)^2} - \frac{u_y' v a_x'/c^2}{\gamma^2(1 + vu_x'/c^2)^3}, \tag{7-2.15}$$

$$a_z = \frac{a_z'}{\gamma^2(1 + vu_x'/c^2)^2} - \frac{u_z' v a_z'/c^2}{\gamma^2(1 + vu_x'/c^2)^3}. \tag{7-2.16}$$

The inverse transformation equations are

$$a_x' = \frac{a_x}{\gamma^3(1 - vu_x/c^2)^3}, \tag{7-2.17}$$

$$a_y' = \frac{a_y}{\gamma^2(1 - vu_x/c^2)^2} + \frac{u_y v a_x/c^2}{\gamma^2(1 - vu_x/c^2)^3}, \tag{7-2.18}$$

$$a_z' = \frac{a_z}{\gamma^2(1 - vu_x/c^2)^2} + \frac{u_z v a_z/c^2}{\gamma^2(1 - vu_x/c^2)^3}. \tag{7-2.19}$$

Let us now obtain a transformation equation for the expression $1 - u'^2/c^2$, which frequently occurs in relativistic calculations.

Consider a charge distribution moving with velocity $\mathbf{u}'$ in the reference frame Σ'. The magnitude of $\mathbf{u}'$ is given by

$$u'^2 = u_x'^2 + u_y'^2 + u_z'^2. \tag{7-2.20}$$

Using Eqs. (7-2.8)-(7-2.10), we can write Eq. (7-2.20) as

$$u'^2 = \frac{(u_x - v)^2 + (u_y^2 + u_z^2)(1 - v^2/c^2)}{(1 - vu_x/c^2)^2}. \tag{7-2.21}$$

For the y and z components of u in Σ we have

$$u_y^2 + u_z^2 = u^2 - u_x^2. \tag{7-2.22}$$

Combining Eqs. (7-2.21) and (7-2.22) and dividing by c^2, we can write for $1 - u'^2/c^2$

$$1 - u'^2/c^2 = 1 - \frac{(u_x/c - v/c)^2 + (u^2/c^2 - u_x^2/c^2)(1 - v^2/c^2)}{(1 - vu_x/c^2)^2}, \tag{7-2.23}$$

which after simplifications becomes

$$1 - u'^2/c^2 = \frac{1 - u^2/c^2}{\gamma^2(1 - vu_x/c^2)^2}. \tag{7-2.24}$$

The inverse transformation equation is, as usual,

$$1 - u^2/c^2 = \frac{1 - u'^2/c^2}{\gamma^2(1 + vu_x'/c^2)^2}. \tag{7-2.25}$$

7-3. Transformation Equations for Partial Derivatives with Respect to Coordinates and Time

We have arrived at the relativistic transformation equations for coordinates, time, fields, and potentials by converting electric and magnetic fields of a moving charge distribution into the corresponding fields of a stationary charge distribution. As we know from Chapter 2, the electric and magnetic field equations that we used for this purpose are solutions of Maxwell's equations. We may suspect therefore that Maxwell's equations themselves can be transformed from one reference frame to another by means of the same transformation equations. We shall explore this possibility in the next section.

Since Maxwell's equations involve partial differentiation with respect to space coordinates as well as partial differentiation with respect to time, we need to know how to transform these operations from one reference frame to another.

Let us first find the equations for transforming $\partial/\partial x$ and $\partial/\partial t$ from the rest frame Σ (laboratory) to the moving frame Σ'. The transformation must take into account that, according to Eqs. (7-1.1) and (7-1.4), a variation of x alone or t alone in Σ is associated with a variation of both x' and t' in Σ', so that for the purpose of the transformation, a function of x or t must be treated as a function of x' and t'.

For $\partial/\partial x$ we then have

$$\frac{\partial}{\partial x} = \frac{\partial}{\partial x'}\frac{\partial x'}{\partial x} + \frac{\partial}{\partial t'}\frac{\partial t'}{\partial x}. \tag{7-3.1}$$

Now, by Eq. (7-1.19), $\partial x'/\partial x = \gamma$, and by Eq. (7-1.22), $\partial t'/\partial x = -\gamma v/c^2$. Therefore Eq. (7-3.1) becomes

$$\frac{\partial}{\partial x} = \gamma\left(\frac{\partial}{\partial x'} - \frac{v}{c^2}\frac{\partial}{\partial t'}\right). \tag{7-3.2}$$

The inverse equation, obtained by transposing the primes and changing the sign in front of v, is

$$\frac{\partial}{\partial x'} = \gamma\left(\frac{\partial}{\partial x} + \frac{v}{c^2}\frac{\partial}{\partial t}\right). \tag{7-3.3}$$

The corresponding hybrid equations are

$$\frac{\partial}{\partial x'} = \frac{1}{\gamma}\frac{\partial}{\partial x} + \frac{v}{c^2}\frac{\partial}{\partial t'} \tag{7-3.4}$$

and

$$\frac{\partial}{\partial x} = \frac{1}{\gamma}\frac{\partial}{\partial x'} - \frac{v}{c^2}\frac{\partial}{\partial t}. \tag{7-3.5}$$

For $\partial/\partial t$ we similarly have

$$\frac{\partial}{\partial t} = \frac{\partial}{\partial x'}\frac{\partial x'}{\partial t} + \frac{\partial}{\partial t'}\frac{\partial t'}{\partial t}. \tag{7-3.6}$$

By Eq. (7-1.19), $\partial x'/\partial t = -\gamma v$, and by Eq. (7-1.22), $\partial t'/\partial t = \gamma$. Therefore Eq. (7-3.6) becomes

$$\frac{\partial}{\partial t} = \gamma\left(\frac{\partial}{\partial t'} - v\frac{\partial}{\partial x'}\right). \tag{7-3.7}$$

The inverse equation is

$$\frac{\partial}{\partial t'} = \gamma\left(\frac{\partial}{\partial t} + v\frac{\partial}{\partial x}\right). \tag{7-3.8}$$

The corresponding hybrid equations are

$$\frac{\partial}{\partial t'} = \frac{1}{\gamma}\frac{\partial}{\partial t} + v\frac{\partial}{\partial x'} \tag{7-3.9}$$

and

$$\frac{\partial}{\partial t} = \frac{1}{\gamma}\frac{\partial}{\partial t'} - v\frac{\partial}{\partial x}. \tag{7-3.10}$$

By Eqs. (7-1.2) and (7-1.3) or (7-1.20) and (7-1.21), the derivatives with respect to y and z transform simply as

$$\frac{\partial}{\partial y} = \frac{\partial}{\partial y'} \tag{7-3.11}$$

and

$$\frac{\partial}{\partial z} = \frac{\partial}{\partial z'}. \tag{7-3.12}$$

7-4. The Invariance of the Cartesian Components of Maxwell's Equations under Relativistic Transformations

The significance of the relativistic transformations presented in the preceding sections of this chapter is twofold: First, the transformations make it possible to correlate electromagnetic quantities measured in different reference frames. Second, as we shall now show, subject to certain limitations to be explained below, Maxwell's equations are invariant with respect to these transformations.[6] Therefore also solutions of Maxwell's equations are invariant with respect to these transformations. This means, among other things, that with the help of relativistic transformations we can obtain solutions to problems involving uniformly moving electromagnetic systems by merely applying relativistic transformations to solutions obtained for the same stationary electromagnetic systems. And since electric and magnetic fields of stationary electromagnetic systems can be easily determined, relativistic transformations provide a powerful and convenient special method for analyzing uniformly moving electromagnetic systems and solving problems pertaining to these systems.

Let us now show that Maxwell's equations (two of them only in their scalar form) are invariant with respect to relativistic transformations. Some special methods based on this invariance will be developed and demonstrated in the next chapter.

***Transformation of* $\nabla\cdot\mathbf{D}=\rho$.** Remembering that $\mathbf{D}=\varepsilon_0\mathbf{E}$ and writing Maxwell's Eq. (2-1.1) in terms of Cartesian components, we have

$$\varepsilon_0\frac{\partial E_x}{\partial x}+\varepsilon_0\frac{\partial E_y}{\partial y}+\varepsilon_0\frac{\partial E_z}{\partial z}=\rho. \tag{7-4.1}$$

Using the hybrid Eq. (7-3.5) and Eq. (7-1.5), using Eq. (7-1.2) and the hybrid Eq. (7-1.37), using Eq. (7-1.3) and the hybrid equation for E_z obtained from Eqs. (7-1.25), and using the hybrid equation for ρ obtained from Eq. (7-1.29), we can write Eq. (7-4.1) as

$$\varepsilon_0\frac{\partial E_x'}{\gamma\partial x'}-\varepsilon_0\frac{v}{c^2}\frac{\partial E_x}{\partial t}+\varepsilon_0\frac{\partial E_y'}{\gamma\partial y'}+\varepsilon_0 v\frac{\partial B_z}{\partial y}+\varepsilon_0\frac{\partial E_z'}{\gamma\partial z'}-\varepsilon_0 v\frac{\partial B_y}{\partial z}$$
$$=\frac{1}{\gamma}\rho'+\frac{v}{c^2}J_x. \tag{7-4.2}$$

Rearranging, we have

$$\begin{aligned}&\frac{1}{\gamma}\left(\frac{\partial\varepsilon_0E_x'}{\partial x'}+\frac{\partial\varepsilon_0E_y'}{\partial y'}+\frac{\partial\varepsilon_0E_z'}{\partial z'}\right)\\&=\frac{1}{\gamma}\rho'-\varepsilon_0v\left(\frac{\partial B_z}{\partial y}-\frac{\partial B_y}{\partial z}\right)+\frac{v}{c^2}\left(J_x+\frac{\partial\varepsilon_0E_x}{\partial t}\right).\end{aligned} \tag{7-4.3}$$

However, since $\mathbf{B} = \mu_0\mathbf{H}$, since $\varepsilon_0\mu_0 = 1/c^2$, and since $\varepsilon_0\mathbf{E} = \mathbf{D}$, the last two terms in Eq. (7-4.3) are simply the x component of the expression

$$\frac{v}{c^2}\left(-\nabla\times\mathbf{H} + \mathbf{J} + \frac{\partial\mathbf{D}}{\partial t}\right), \tag{7-4.4}$$

which by Maxwell's Eq. (2-1.4) is zero. Hence, dropping the last two terms in Eq. (7-4.3), cancelling γ, replacing $\varepsilon_0\mathbf{E}'$ by $\mathbf{D}'$, and restoring the vector notation, we obtain

$$\nabla'\cdot\mathbf{D}' = \rho'. \tag{7-4.5}$$

Thus Maxwell's Eq. (2-1.1) is invariant under relativistic transformations.

Transformation of $\nabla\cdot\mathbf{B} = 0$. Writing Maxwell's Eq. (2-1.2) in terms of Cartesian components, we have

$$\frac{\partial B_x}{\partial x} + \frac{\partial B_y}{\partial y} + \frac{\partial B_z}{\partial z} = 0. \tag{7-4.6}$$

Using the hybrid Eq. (7-3.5) and Eq. (7-1.8), using Eqs. (7-1.2), (7-1.3) and the hybrid equations for B_y and B_z obtained from Eqs. (7-1.27) and (7-1.28), we can write Eq. (7-4.6) as

$$\frac{\partial B_x'}{\gamma\partial x'} - \frac{v}{c^2}\frac{\partial B_x}{\partial t} + \frac{\partial B_y'}{\gamma\partial y'} - \frac{v}{c^2}\frac{\partial E_z}{\partial y} + \frac{\partial B_z'}{\gamma\partial z'} + \frac{v}{c^2}\frac{\partial E_y}{\partial z} = 0. \tag{7-4.7}$$

Multiplying by γ and rearranging, we have

$$\frac{\partial B_x'}{\partial x'} + \frac{\partial B_y'}{\partial y'} + \frac{\partial B_z'}{\partial z'} = \gamma\frac{v}{c^2}\left[\left(\frac{\partial E_z}{\partial y} - \frac{\partial E_y}{\partial z}\right) + \frac{\partial B_x}{\partial t}\right]. \tag{7-4.8}$$

However, the expression in the brackets is simply the x component of the expression

$$\frac{v}{c^2}\left(\nabla\times\mathbf{E} + \frac{\partial\mathbf{B}}{\partial t}\right), \tag{7-4.9}$$

which by Maxwell's Eq. (2-1.3) is zero. Replacing the right side of Eq. (7-4.8) by zero and restoring the vector notation, we obtain

$$\nabla' \cdot \mathrm{B}' = 0. \tag{7-4.10}$$

Thus Maxwell's Eq. (2-1.2) is invariant under relativistic transformations.

Transformation of $\nabla \times \mathbf{E} = -\partial\mathbf{B}/\partial t$. Writing Maxwell's Eq. (2-1.3) in terms of Cartesian components, we have

$$\begin{aligned}\mathbf{i}\left(\frac{\partial E_z}{\partial y} - \frac{\partial E_y}{\partial z}\right) + \mathbf{j}\left(\frac{\partial E_x}{\partial z} - \frac{\partial E_z}{\partial x}\right) + \mathbf{k}\left(\frac{\partial E_y}{\partial x} - \frac{\partial E_x}{\partial y}\right)\\ = -\mathbf{i}\frac{\partial B_x}{\partial t} - \mathbf{j}\frac{\partial B_y}{\partial t} - \mathbf{k}\frac{\partial B_z}{\partial t}.\end{aligned} \tag{7-4.11}$$

Using Eqs. (7-1.2), (7-1.3), (7-1.5)-(7-1.8), and (7-3.7), we can write Eq. (7-4.11) as

$$\begin{aligned}&\mathbf{i}\left(\gamma\frac{\partial E_z'}{\partial y'} - \gamma v\frac{\partial B_y'}{\partial y'} - \gamma\frac{\partial E_y'}{\partial z'} - \gamma v\frac{\partial B_z'}{\partial z'}\right)\\ &+ \mathbf{j}\left(\frac{\partial E_x'}{\partial z'} - \frac{\partial E_z}{\partial x}\right) + \mathbf{k}\left(\frac{\partial E_y}{\partial x} - \frac{\partial E_x'}{\partial y'}\right)\\ &= -\mathbf{i}\left(\gamma\frac{\partial B_x'}{\partial t'} - \gamma v\frac{\partial B_x'}{\partial x'}\right) - \mathbf{j}\frac{\partial B_y}{\partial t} - \mathbf{k}\frac{\partial B_z}{\partial t}.\end{aligned} \tag{7-4.12}$$

According to Eq. (7-4.10), the terms with the derivatives $\partial B'_x/\partial x'$, $\partial B'_y/\partial y'$, $\partial B'_z/\partial z'$ in Eq. (7-4.12) vanish, so that the equation simplifies to

$$\begin{aligned}\mathbf{i}\left(\gamma\frac{\partial E_z'}{\partial y'} - \gamma\frac{\partial E_y'}{\partial z'}\right) + \mathbf{j}\left(\frac{\partial E_x'}{\partial z'} - \frac{\partial E_z}{\partial x}\right) + \mathbf{k}\left(\frac{\partial E_y}{\partial x} - \frac{\partial E_x'}{\partial y'}\right)\\ = -\mathbf{i}\gamma\frac{\partial B_x'}{\partial t'} - \mathbf{j}\frac{\partial B_y}{\partial t} - \mathbf{k}\frac{\partial B_z}{\partial t}.\end{aligned} \tag{7-4.13}$$

Using Eqs. (7-1.7), (7-1.6), (7-1.9), and (7-1.10), we can write

Eq. (7-4.13) as

$$\mathbf{i}\left(\gamma\frac{\partial E_z'}{\partial y'} - \gamma\frac{\partial E_y'}{\partial z'}\right) + \mathbf{j}\left[\frac{\partial E_x'}{\partial z'} - \gamma\frac{\partial(E_z' - vB_y')}{\partial x}\right] + \mathbf{k}\left[\gamma\frac{\partial(E_y + vB_z')}{\partial x} - \frac{\partial E_x'}{\partial y'}\right]$$
$$= -\mathbf{i}\gamma\frac{\partial B_x'}{\partial t'} - \mathbf{j}\gamma\frac{\partial(B_y' - vE_z'/c^2)}{\partial t} - \mathbf{k}\gamma\frac{\partial(B_z' + vE_y'/c^2)}{\partial t} \tag{7-4.14}$$

or, rearranging, as

$$\mathbf{i}\left(\gamma\frac{\partial E_z'}{\partial y'} - \gamma\frac{\partial E_y'}{\partial z'}\right)$$
$$+ \mathbf{j}\left[\frac{\partial E_x'}{\partial z'} - \gamma\left(\frac{\partial E_z'}{\partial x} + \frac{v}{c^2}\frac{\partial E_z'}{\partial t}\right)\right] + \mathbf{k}\left[\gamma\left(\frac{\partial E_y'}{\partial x} + \frac{v}{c^2}\frac{\partial E_y'}{\partial t}\right) - \frac{\partial E_x'}{\partial y'}\right]$$
$$= -\mathbf{i}\gamma\frac{\partial B_x'}{\partial t'} - \mathbf{j}\gamma\left(\frac{\partial B_y'}{\partial t} + v\frac{\partial B_y'}{\partial x}\right) - \mathbf{k}\gamma\left(\frac{\partial B_z'}{\partial t} + v\frac{\partial B_z'}{\partial x}\right), \tag{7-4.15}$$

which, by Eqs. (7-3.3) and (7-3.8), is

$$\mathbf{i}\left(\gamma\frac{\partial E_z'}{\partial y'} - \gamma\frac{\partial E_y'}{\partial z'}\right) + \mathbf{j}\left(\frac{\partial E_x'}{\partial z'} - \frac{\partial E_z'}{\partial x'}\right) + \mathbf{k}\left(\frac{\partial E_y'}{\partial x'} - \frac{\partial E_x'}{\partial y'}\right)$$
$$= -\mathbf{i}\gamma\frac{\partial B_x'}{\partial t'} - \mathbf{j}\frac{\partial B_y'}{\partial t'} - \mathbf{k}\frac{\partial B_z'}{\partial t'}. \tag{7-4.16}$$

Comparing the x, y, and z components of the left side of Eq. (7-4.16) with those of the right side, we find that the components have the same form as the components of Eq. (7-4.11) (the factor γ in the x components cancels if one compares only the individual components of the left and the right side of the equation). Thus the *Cartesian components* of Maxwell's Eq. (2-1.3) are invariant under relativistic transformations, but the equation itself is not invariant because, due to the presence of γ in the x components of Eq. (7-4.16), Eq. (7-4.16) is not the same as Eq. (7-4.11).

Transformation of $\nabla \times \mathbf{H} = \mathbf{J} + \partial\mathbf{D}/\partial t$. Remembering that $\mathbf{D} = \varepsilon_0\mathbf{E}$ and writing Maxwell's Eq. (2-1.4) in terms of Cartesian components, we have

$$\mathbf{i}\left(\frac{\partial H_z}{\partial y} - \frac{\partial H_y}{\partial z}\right) + \mathbf{j}\left(\frac{\partial H_x}{\partial z} - \frac{\partial H_z}{\partial x}\right) + \mathbf{k}\left(\frac{\partial H_y}{\partial x} - \frac{\partial H_x}{\partial y}\right)$$
$$= \mathbf{i}J_x + \mathbf{j}J_y + \mathbf{k}J_z + \mathbf{i}\varepsilon_0\frac{\partial E_x}{\partial t} + \mathbf{j}\varepsilon_0\frac{\partial E_y}{\partial t} + \mathbf{k}\varepsilon_0\frac{\partial E_z}{\partial t}. \tag{7-4.17}$$

Using Eqs. (7-1.2), (7-1.3), (7-1.5), (7-1.8)-(7-1.10), (7-1.12)-(7-1.14), and (7-3.7), and remembering that $\mathbf{B} = \mu_0\mathbf{H}$, we can write Eq. (7-4.17) as

$$\mathbf{i}\left(\gamma\frac{\partial H_z'}{\partial y'} + \gamma v\frac{\partial E_y'}{\mu_0 c^2\partial y'} - \gamma\frac{\partial H_y'}{\partial z'} + \gamma v\frac{\partial E_z'}{\mu_0 c^2 \partial z'}\right)$$
$$+\mathbf{j}\left(\frac{\partial H_x'}{\partial z'} - \frac{\partial H_z}{\partial x}\right) + \mathbf{k}\left(\frac{\partial H_y}{\partial x} - \frac{\partial H_x'}{\partial y'}\right) \tag{7-4.18}$$
$$= \mathbf{i}\gamma(J_x' + v\rho') + \mathbf{j}J_y' + \mathbf{k}J_z' + \mathbf{i}\varepsilon_0\gamma\left(\frac{\partial E_x'}{\partial t'} - v\frac{\partial E_x'}{\partial x'}\right) + \mathbf{j}\varepsilon_0\frac{\partial E_y}{\partial t} + \mathbf{k}\varepsilon_0\frac{\partial E_z}{\partial t}.$$

According to Eq. (7-4.5) and taking into account that $1/\mu_0 c^2 = \varepsilon_0$, the terms with the derivatives $\partial E'_x/\partial x'$, $\partial E'_y/\partial y'$, $\partial E'_z/\partial z'$ and ρ' in Eq. (7-4.18) vanish, so that the equation simplifies to

$$\mathbf{i}\left(\gamma\frac{\partial H_z'}{\partial y'} - \gamma\frac{\partial H_y'}{\partial z'}\right) + \mathbf{j}\left(\frac{\partial H_x'}{\partial z'} - \frac{\partial H_z}{\partial x}\right) + \mathbf{k}\left(\frac{\partial H_y}{\partial x} - \frac{\partial H_x'}{\partial y'}\right)$$
$$= \mathbf{i}\gamma J_x' + \mathbf{j}J_y' + \mathbf{k}J_z' + \mathbf{i}\gamma\varepsilon_0\frac{\partial E_x'}{\partial t'} + \mathbf{j}\varepsilon_0\frac{\partial E_y}{\partial t} + \mathbf{k}\varepsilon_0\frac{\partial E_z}{\partial t}. \tag{7-4.19}$$

Using Eqs. (7-1.6), (7-1.7), (7-1.9), and (7-1.10), we can write Eq. (7-4.19) as

$$\mathbf{i}\gamma\left(\frac{\partial H_z'}{\partial y'} - \frac{\partial H_y'}{\partial z'}\right)$$
$$+\mathbf{j}\left[\frac{\partial H_x'}{\partial z'} - \gamma\frac{\partial(H_z' + vE_y'/\mu_0 c^2)}{\partial x}\right] + \mathbf{k}\left[\gamma\frac{\partial(H_y' - vE_z'/\mu_0 c^2)}{\partial x} - \frac{\partial H_x'}{\partial y'}\right]$$
$$= \mathbf{i}\gamma J_x' + \mathbf{j}J_y' + \mathbf{k}J_z' + \mathbf{i}\gamma\varepsilon_0\frac{\partial E_x'}{\partial t'} + \mathbf{j}\varepsilon_0\gamma\frac{\partial(E_y' + vB_z')}{\partial t} + \mathbf{k}\varepsilon_0\gamma\frac{\partial(E_z' - vB_y')}{\partial t} \tag{7-4.20}$$

or, noting that $\varepsilon_0 B = H/c^2$, $1/\mu_0 c^2 = \varepsilon_0$ and rearranging, as

$$\mathbf{i}\left(\gamma\frac{\partial H_z'}{\partial y'} - \gamma\frac{\partial H_y'}{\partial z'}\right)$$
$$+\mathbf{j}\left[\frac{\partial H_x'}{\partial z'} - \gamma\left(\frac{\partial H_z'}{\partial x} + \frac{v}{c^2}\frac{\partial H_z'}{\partial t}\right)\right] + \mathbf{k}\left[\gamma\left(\frac{\partial H_y'}{\partial x} + \frac{v}{c^2}\frac{\partial H_y'}{\partial t}\right) - \frac{\partial H_x'}{\partial y'}\right]$$
$$= \mathbf{i}\gamma J_x' + \mathbf{j}J_y' + \mathbf{k}J_z' + \mathbf{i}\gamma\varepsilon_0\frac{\partial E_x'}{\partial t'}$$
$$+\mathbf{j}\gamma\varepsilon_0\left(\frac{\partial E_y'}{\partial t} + v\frac{\partial E_y'}{\partial x}\right) + \mathbf{k}\gamma\varepsilon_0\left(\frac{\partial E_z'}{\partial t} + v\frac{\partial E_z'}{\partial x}\right), \tag{7-4.21}$$

which, by Eqs. (7-3.3) and (7-3.8), is

$$\mathbf{i}\gamma\left(\frac{\partial H_z'}{\partial y'} - \frac{\partial H_y'}{\partial z'}\right) + \mathbf{j}\left(\frac{\partial H_x'}{\partial z'} - \frac{\partial H_z'}{\partial x'}\right) + \mathbf{k}\left(\frac{\partial H_y'}{\partial x'} - \frac{\partial H_x'}{\partial y'}\right)$$
$$= \mathbf{i}\gamma J_x' + \mathbf{j}J_y' + \mathbf{k}J_z' + \mathbf{i}\gamma\varepsilon_0\frac{\partial E_x'}{\partial t'} + \mathbf{j}\varepsilon_0\frac{\partial E_y'}{\partial t'} + \mathbf{k}\varepsilon_0\frac{\partial E_z'}{\partial t'}. \tag{7-4.22}$$

Comparing the x, y, and z components of the left side of Eq. (7-4.22) with those of the right side, we find that the components have the same form as the components of Eq. (7-4.17) (the factor γ in the x components cancels if one equates only the individual components of the left and the right side of the equation). Thus the *Cartesian components* of Maxwell's Eq. (2-1.4) are invariant under relativistic transformations, but the equation itself is not invariant because, due to the presence of γ in the x components of Eq. (7-4.22), Eq. (7-4.22) is not the same as Eq. (7-4.17).[7]

7-5. Testing Relativistic Transformations

Although we have no reason to doubt the correctness of our derivations and the correctness of the relativistic transformations that we have obtained, it is instructive to test some of the transformation equations. We can do so by using relativistic transformations for solving some problems whose solution is already known on the basis of general electromagnetic laws.

Correlation between electric and magnetic fields of a moving charge distribution. For the first test, let us see what effect relativistic transformations have on the relation between the electric and magnetic fields of a moving point charge. Consider the equation expressing the magnetic flux density field **B** of a uniformly moving charge distribution in terms of the electric field **E** and the velocity **v** of the distribution [Eq. (3-2.10)]

$$\mathbf{B} = (\mathbf{v} \times \mathbf{E})/c^2. \tag{7-5.1}$$

By the relativity principle, this equation should not depend on the reference frame in which **E**, **v**, and **B** are measured. Let us see if this conclusion is supported by our transformation equations.

Let a charge distribution move with velocity $\mathbf{u}'$ with respect to a reference frame Σ', which moves with velocity $\mathbf{v} = v\mathbf{i}$ with respect to the laboratory. In Σ' Eq. (7-5.1) is then

$$\mathbf{B}' = (\mathbf{u}' \times \mathbf{E}')/c^2. \tag{7-5.2}$$

We shall now transform this equation to the laboratory frame. To do so we first write Eq. (7-5.2) in terms of its Cartesian components

$$B_x' = (u_y'E_z' - u_z'E_y')/c^2 \tag{7-5.3}$$

$$B_y' = (u_z'E_x' - u_x'E_z')/c^2 \tag{7-5.4}$$

$$B_z' = (u_x'E_y' - u_y'E_x')/c^2. \tag{7-5.5}$$

Substituting into Eq. (7-5.5) E_x' from Eq. (7-1.5) and the hybrid equations for E_y', and E_z' obtained from Eq. (7-1.6) and (7-1.7), we have

$$B_x' = [u_y'(E_z/\gamma + vB_y') - u_z'(E_y/\gamma - vB_z')]/c^2 \tag{7-5.6}$$

$$B_y' = [u_z'E_x - u_x'(E_z/\gamma + vB_y')]/c^2 \tag{7-5.7}$$

$$B_z' = [u_x'(E_y/\gamma - vB_z') - u_y'E_x]/c^2. \tag{7-5.8}$$

We shall now simplify Eq. (7-5.6) with the help of Eq. (7-5.2) by using the relation

$$\mathbf{u}' \cdot \mathrm{B}' = \mathbf{u}' \cdot (\mathbf{u}' \times \mathbf{E}')/c^2 = 0, \tag{7-5.9}$$

from which it follows that

$$u_y'B_y' + u_z'B_z' = - u_x'B_x'. \tag{7-5.10}$$

Substituting Eq. (7-5.10) into Eq. (7-5.6), we obtain

$$B_x' = (u_y'E_z/\gamma - u_z'E_y/\gamma - vu_x'B_x')/c^2 \tag{7-5.11}$$

or

$$B_x'(1 + vu_x'/c^2) = (u_y'E_z/\gamma - u_z'E_y/\gamma)/c^2, \tag{7-5.12}$$

so that

$$B_x' = \left[\frac{u_y'}{\gamma(1 + vu_x'/c^2)}E_z - \frac{u_z'}{\gamma(1 + vu_x'/c^2)}E_y\right]\frac{1}{c^2}, \tag{7-5.13}$$

which, by Eqs. (7-1.26), (7-2.6), and (7-2.7), is

$$B_x = (u_yE_z - u_zE_y)/c^2. \tag{7-5.14}$$

Rearranging now Eq. (7-5.7), we have

$$B_y'(1 + vu_x'/c^2) = (u_z'E_x - u_x'E_z/\gamma)/c^2 \tag{7-5.15}$$

or

$$B_y' = \left[\frac{u_z'}{(1 + vu_x'/c^2)}E_x - \frac{u_x'}{\gamma(1 + vu_x'/c^2)}E_z\right]\frac{1}{c^2}. \tag{7-5.16}$$

Substituting B_y' from Eq. (7-1.27), we obtain

$$\gamma(B_y + vE_z/c^2) = \left[\frac{u_z}{(1 + vu_x'/c^2)}E_x - \frac{u_x'}{\gamma(1 + vu_x'/c^2)}E_z\right]\frac{1}{c^2} \tag{7-5.17}$$

or

$$\begin{aligned} B_y &= \left[\frac{u_z'}{\gamma(1 + vu_x'/c^2)}E_x - \frac{u_x'}{\gamma^2(1 + vu_x'/c^2)}E_z - vE_x\right]\frac{1}{c^2} \\ &= \left[\frac{u_z'}{\gamma(1 + vu_x'/c^2)}E_x - \frac{u_x'(1 - v^2/c^2) + v(1 + vu_x'/c^2)}{(1 + vu_x'/c^2)}E_z\right]\frac{1}{c^2} \\ &= \left[\frac{u_z'}{\gamma(1 + vu_x'/c^2)}E_x - \frac{u_x' + v}{(1 + vu_x'/c^2)}E_z\right]\frac{1}{c^2}, \end{aligned} \tag{7-5.18}$$

which, by Eqs. (7-2.7) and (7-2.5), is

$$B_y = (u_zE_x - u_xE_z)/c^2. \tag{7-5.19}$$

Clearly, Eq. (7-.5.5) transforms in the same manner into

$$B_z = (u_xE_y - u_yE_x)/c^2. \tag{7-5.20}$$

Recombining Eqs. (7-5.14), (7-5.19), and (7-5.20) into a single vector equation, we finally obtain Eq. (7-5.1) thus demonstrating the validity of our transformations.

Electric field of a moving point charge. For the second test, let us see what effect relativistic transformations have on the electric field of a moving point charge. Consider a point charge q moving with constant velocity $\mathbf{u}' = u'\mathbf{i}$ relative to a reference frame Σ', which moves with velocity $\mathbf{v} = v\mathbf{i}$ relative to the laboratory (reference frame Σ). Let the charge be in the $x'y'$ plane, let the point of observation in Σ' be at $x' = 0$, $y' = 0$, $z' = 0$, let the time of observation in Σ' be $t' = 0$, and let the point of observation in Σ be at $x = 0$, $y = 0$, $z = 0$. As usual, let the x' axis coincide with the x axis, and let the $x'y'$ plane coincide with the xy plane.

The electric field produced by q in Σ' is, by Eq. (4-1.19),

$$\mathbf{E}' = \frac{q(1 - u'^2/c^2)(x_0'\mathbf{i} + y'\mathbf{j})}{4\pi\varepsilon_0\{x_0'^2 + (1 - u'^2/c^2)y'^2\}^{3/2}}, \tag{7-5.21}$$

where x_0' is the x' coordinate of the point charge at $t' = 0$. If our relativistic transformation equations are correct, then the only effect of these transformations on Eq. (7-5.21) when the equation is transformed to the reference frame Σ should be the absence of the primes in the equation.

To perform the transformation, let us first write the equation in terms of its Cartesian components. We have

$$E_x' = \frac{q(1 - u'^2/c^2)x_0'}{4\pi\varepsilon_0\{x_0'^2 + (1 - u'^2/c^2)y'^2\}^{3/2}} \tag{7-5.22}$$

and

$$E_y' = \frac{q(1 - u'^2/c^2)y'}{4\pi\varepsilon_0\{x_0'^2 + (1 - u'^2/c^2)y'^2\}^{3/2}}. \tag{7-5.23}$$

Substituting now into Eq. (7-5.22) Eqs. (7-1.23), (7-2.24), (7-1.20), and the hybrid equation for x' obtained from Eq. (7-1.1) with $t' = 0$, and noting that in the case under consideration $u_x' = u'$, we obtain

$$E_x = \frac{q(1-u^2/c^2)x/\gamma}{4\pi\varepsilon_0\gamma^2(1-vu/c^2)^2\{x^2/\gamma^2+(1-u^2/c^2)/\gamma^2(1-vu/c^2)^2y^2\}^{3/2}}$$

$$= \frac{q(1-u^2/c^2)(1-vu/c^2)^2x}{4\pi\varepsilon_0\{x^2(1-vu/c^2)^2+(1-u^2/c^2)y^2\}^{3/2}}. \tag{7-5.24}$$

Now, since in Σ' the charge was observed at $t' = 0$, x in Eq. (7-5.24) is, according to Eq. (7-1.22), the position of the charge at $t = vx/c^2$. But to make the electric field given by Eq. (7-5.24) correspond to the field observed in Σ', the time of observation in Σ must be the same as in Σ', that is, $t = 0$. Therefore we must replace x in Eq. (7-5.24) by x_0, the position occupied by the charge at $t = 0$. Setting

$$x = x_0 + ut = x_0 + u(vx/c^2) \tag{7-5.25}$$

and solving for x, we obtain

$$x = \frac{x_0}{1 - vu/c^2}. \tag{7-5.26}$$

Substituting Eq. (7-5.26) into Eq. (7-5.24), we obtain

$$E_x = \frac{q(1-u^2/c^2)x_0}{4\pi\varepsilon_0\{x_0^2+(1-u^2/c^2)y^2\}^{3/2}}. \tag{7-5.27}$$

For transforming Eq. (7-5.23), we need to use Eq. (7-1.24) which contains B_z. To obtain B_z, we use Eq. (7-5.1), which gives (note that the velocity of the charge in Σ is u)

$$B_z = uE_y/c^2. \tag{7-5.28}$$

Substituting Eq. (7-5.28) into Eq. (7-1.24), we obtain

$$E_y' = \gamma E_y(1 - vu/c^2). \tag{7-5.29}$$

Substituting now into Eq. (7-5.23) Eqs. (7-5.29), (7-2.24), (7-1.20), and the hybrid equation for x' obtained from Eq. (7-1.1) with $t' = 0$, and taking into account that in the case under consideration $u_x' = u'$, we obtain

$$\gamma E_y(1 - vu/c^2)$$
$$= \frac{q(1 - u^2/c^2)y}{4\pi\varepsilon_0\gamma^2(1 - vu/c^2)^2\{x^2/\gamma^2 + (1 - u^2/c^2)/\gamma^2(1 - vu/c^2)^2 y^2\}^{3/2}}$$
$$= \frac{q(1 - u^2/c^2)(1 - vu/c^2)^2\gamma y}{4\pi\varepsilon_0\{x^2(1 - vu/c^2)^2 + (1 - u^2/c^2)y^2\}^{3/2}} \tag{7-5.30}$$

or

$$E_y = \frac{q(1 - u^2/c^2)y}{4\pi\varepsilon_0\{x^2(1 - vu/c^2)^2 + (1 - u^2/c^2)y^2\}^{3/2}}. \tag{7-5.31}$$

Substituting Eq. (7-5.26) into Eq. (7-5.31), we obtain

$$E_y = \frac{q(1 - u^2/c^2)y}{4\pi\varepsilon_0\{x_0^2 + (1 - u^2/c^2)y^2\}^{3/2}}. \tag{7-5.32}$$

Recombining Eqs. (7-5.32) and (7-5.27) into a single vector equation, we finally obtain

$$\mathbf{E} = \frac{q(1 - u^2/c^2)(x_0\mathbf{i} + y\mathbf{j})}{4\pi\varepsilon_0\{x_0^2 + (1 - u^2/c^2)y^2\}^{3/2}}, \tag{7-5.33}$$

thus once again demonstrating the validity of our transformations.[8]

7-6. The Method of Corresponding States

In 1895, H. A. Lorentz enunciated a theorem, which he called the *theorem of corresponding states*, according to which to any electromagnetic system that is a function of space and time coordinates in the rest frame Σ, there corresponds an electromagnetic system in the moving frame Σ', being the same function of space and time coordinates (primed coordinates) in Σ'.[9] The theorem constitutes one of the most effective tools of relativistic electrodynamics, making possible a very simple derivation of various equations for electric and magnetic fields of uniformly moving charge distributions from the corresponding electrostatic and magnetostatic equations. Several examples of the use of this theorem are provided below.[10]

▼

Example 7-6.1 The electric field of a stationary charge distribution can be found from[11]

$$\mathbf{E} = -\frac{1}{4\pi\varepsilon_0}\int \frac{\nabla\rho}{r}dV. \tag{7-6.1}$$

Using Eq. (7-6.1) and appropriate transformation equations, find the electric field produced by a charge distribution moving with uniform velocity $\mathbf{v} = v\mathbf{i}$.

Let us apply Eq. (7-6.1) to a charge distribution ρ' resting in a reference frame Σ' which moves with respect to the laboratory (reference frame Σ) with constant velocity $\mathbf{v} = v\mathbf{i}$. The Cartesian components of the electric field $\mathbf{E}'$ produced by ρ' in this reference frame are the same as those of Eq. (7-6.1) with ∇, ρ, $\mathbf{r}$, and dV replaced by the corresponding primed quantities, that is

$$E'_x = -\frac{1}{4\pi\varepsilon_0}\int \frac{(\partial/\partial x')\rho'}{(x'^2+y'^2+z'^2)^{1/2}}dx'dy'dz', \tag{7-6.2}$$

$$E'_y = -\frac{1}{4\pi\varepsilon_0}\int \frac{(\partial/\partial y')\rho'}{(x'^2+y'^2+z'^2)^{1/2}}dx'dy'dz', \tag{7-6.3}$$

$$E'_z = -\frac{1}{4\pi\varepsilon_0}\int \frac{(\partial/\partial z')\rho'}{(x'^2+y'^2+z'^2)^{1/2}}dx'dy'dz'. \tag{7-6.4}$$

Since the electric field in Σ' does not depend on time, we are free to choose the time of observation t' in Σ' and therefore, by Eq. (7-1.4), the time of observation t in Σ. For simplicity we shall use $t = 0$. Also since the electric field in Σ' does not depend on time, so that $\partial/\partial t' = 0$, the derivative $\partial/\partial x'$, by Eq. (7-3.22), transforms into $\partial/\gamma\partial x$. Taking into account that there is no magnetic field in Σ' (because the charge distribution is at rest there) and using Eqs. (7-1.5)-(7-1.7), (7-1.19) with $t = 0$, (7-1.20), (7-1.21), (7-6.2), and (7-1.11) with $J_x' = 0$ (because there is no current in Σ') we transform Eqs. (7-6.2)-(7-6.4) into

$$E_x = -\frac{1}{4\pi\varepsilon_0}\int\frac{(\partial/\gamma\partial x)\rho/\gamma}{[(\gamma x)^2+y^2+z^2]^{1/2}}d(\gamma x)dydz, \tag{7-6.5}$$

$$E_y/\gamma = -\frac{1}{4\pi\varepsilon_0}\int\frac{(\partial/\partial y)\rho/\gamma}{[(\gamma x)^2+y^2+z^2]^{1/2}}d(\gamma x)dydz, \tag{7-6.6}$$

$$E_z/\gamma = -\frac{1}{4\pi\varepsilon_0}\int\frac{(\partial/\partial z)\rho/\gamma}{[(\gamma x)^2+y^2+z^2]^{1/2}}d(\gamma x)dydz, \tag{7-6.7}$$

or

$$E_x = -\frac{1}{4\pi\varepsilon_0\gamma^2}\int\frac{(\partial/\partial x)\rho}{[x^2+(y^2+z^2)/\gamma^2]^{1/2}}dV, \tag{7-6.8}$$

$$E_y = -\frac{1}{4\pi\varepsilon_0}\int\frac{(\partial/\partial y)\rho}{[x^2+(y^2+z^2)/\gamma^2]^{1/2}}dV, \tag{7-6.9}$$

$$E_z = -\frac{1}{4\pi\varepsilon_0}\int\frac{(\partial/\partial z)\rho}{[x^2+(y^2+z^2)/\gamma^2]^{1/2}}dV. \tag{7-6.10}$$

The denominators in Eqs. (7-6.8)-(7-6.10) can be simplified with the help of Eqs. (5-1.8) and (5-1.9). Multiplying Eqs. (7-6.8)-(7-6.10) by **i**, **j**, and **k**, respectively, adding the equations, and observing that $1/\gamma^2 = 1 - v^2/c^2$, we obtain

$$\mathbf{E} = -\frac{1}{4\pi\varepsilon_0}\int\frac{\nabla\rho - \mathbf{i}(v^2/c^2)(\partial/\partial x)\rho}{r[1-(v^2/c^2)\sin^2\theta]^{1/2}}dV. \tag{7-6.11}$$

Observe that, except for notation, Eq. (7-6.11) is the same as Eq. (5-1.12) that we obtained by converting the retarded integral for the electric field given by Eq. (5-1.1) into the present-position integral [the primes in Eq. (5-1.12) were used to indicate the source-point coordinates; in Eq. (7-6.11) these coordinates appear without primes, because in relativistic electrodynamics the primes are used for identifying coordinates in moving reference frames].

Example 7-6.2 The best known expression for calculating electric fields of stationary charges is the "Coulomb's field" equation[12]

$$\mathbf{E} = \frac{1}{4\pi\varepsilon_0}\int \frac{\rho\mathbf{r}}{r^3}dV. \tag{7-6.12}$$

Convert this equation into the equation for the electric field produced by a charge distribution moving with constant velocity and obtain the corresponding equation for the magnetic field produced by the moving charge distribution.

Consider a charge distribution ρ' resting in a reference frame Σ' which moves, as usual, with respect to the laboratory (reference frame Σ) with constant velocity $\mathbf{v} = v\mathbf{i}$. The electric field $\mathbf{E}'$ produced by ρ' in Σ' may be found from Eq. (7-6.12). Let us rewrite this equation in terms of its Cartesian components (using primed coordinates, since the coordinates are in Σ')

$$E_x' = \frac{1}{4\pi\varepsilon_0}\int \frac{\rho' x'}{(x'^2+y'^2+z'^2)^{3/2}}dx'dy'dz', \tag{7-6.13}$$

$$E_y' = \frac{1}{4\pi\varepsilon_0}\int \frac{\rho' y'}{(x'^2+y'^2+z'^2)^{3/2}}dx'dy'dz', \tag{7-6.14}$$

$$E_z' = \frac{1}{4\pi\varepsilon_0}\int \frac{\rho' z'}{(x'^2+y'^2+z'^2)^{3/2}}dx'dy'dz'. \tag{7-6.15}$$

To find the electric and magnetic fields that the charge distribution produces in the laboratory, we shall apply to Eqs. (7-6.13)-(7-6.15) our relativistic transformation equations. Since the electric field in the moving reference frame Σ' does not depend on time, we shall use, for simplicity, $t = 0$ for the time of observation in the stationary reference frame Σ. Taking into account that there is no magnetic field in Σ' (because the charge distribution is at rest there) and using Eqs. (7-1.5)-(7-1.7), (7-1.19)-(7-1.21), and (7-1.11) with $J_x' = 0$ (because there is no current in Σ') we transform Eqs. (7-6.13)-(7-6.15) into

$$E_x = \frac{1}{4\pi\varepsilon_0}\int \frac{(\rho/\gamma)\gamma x}{(\gamma^2 x^2 + y^2 + z^2)^{3/2}} d(\gamma x)dydz \qquad (7\text{-}6.16)$$

or

$$E_x = \frac{1}{4\pi\varepsilon_0\gamma^2}\int \frac{\rho x}{[x^2 + (y^2 + z^2)/\gamma^2]^{3/2}} dV, \qquad (7\text{-}6.17)$$

$$\frac{E_y}{\gamma} = \frac{1}{4\pi\varepsilon_0}\int \frac{(\rho/\gamma)y}{(\gamma^2 x^2 + y^2 + z^2)^{3/2}} d(\gamma x)dydz \qquad (7\text{-}6.18)$$

or

$$E_y = \frac{1}{4\pi\varepsilon_0\gamma^2}\int \frac{\rho y}{[x^2 + (y^2 + z^2)/\gamma^2]^{3/2}} dV, \qquad (7\text{-}6.19)$$

and, similarly,

$$E_z = \frac{1}{4\pi\varepsilon_0\gamma^2}\int \frac{\rho z}{[x^2 + (y^2 + z^2)/\gamma^2]^{3/2}} dV. \qquad (7\text{-}6.20)$$

The denominators in Eqs. (7-6.17), (7-6.19), and (7-6.20) can be simplified with the help of Eqs. (5-1.8) and (5-1.9). Recombining Eqs. (7-6.17), (7-6.19), and (7-6.20) into a single vector equation, we then obtain

$$\mathbf{E} = \frac{1}{4\pi\varepsilon_0\gamma^2}\int \frac{\rho \mathbf{r}}{r^3[1 - (v^2/c^2)\sin^2\theta]^{3/2}} dV. \qquad (7\text{-}6.21)$$

Observe that, except for notation, Eq. (7-6.21) is the same as Eq. (5-1.40) which we obtained by converting retarded integrals into present-position integrals.

Although the magnetic field produced by this charge distribution could be found by applying relativistic transformations [Eqs. (7-1.8)-(7-1.10) in particular] to Eqs. (7-6.13)-(7-6.15), it is much easier to find it by applying Eq. (3-2.6) to Eq. (7-6.21). Clearly, this would yield Eq. (5-1.45).

Example 7-6.3 The electric field at a distance R from a *stationary* line charge with endpoints at $x_1 = L_1$ and $x_2 = L_2$, as in Fig. 4.5, is

$$E_x = \frac{\lambda}{4\pi\varepsilon_0 R}\left[\frac{1}{(L_1^2/R^2+1)^{1/2}} - \frac{1}{(L_2^2/R^2+1)^{1/2}}\right], \tag{7-6.22}$$

$$E_y = -\frac{\lambda}{4\pi\varepsilon_0 R^2}\left[\frac{L_1}{(L_1^2/R^2+1)^{1/2}} - \frac{L_2}{(L_2^2/R^2+1)^{1/2}}\right], \tag{7-6.23}$$

where λ is the line density of the charge, and where the point of observation is at the origin.[13] What is the electric field of this line charge if the charge moves parallel to the x axis?

Let us suppose that the charge is at rest in a reference frame Σ' which is moving with velocity $\mathbf{v} = v\mathbf{i}$ relative to the laboratory reference frame Σ along their common x axis. In the Σ' frame the x component of the electric field of the line charge is, by Eq. (7-6.22),

$$E_x' = \frac{\lambda'}{4\pi\varepsilon_0 R'}\left[\frac{1}{(L_1'^2/R'^2+1)^{1/2}} - \frac{1}{(L_2'^2/R'^2+1)^{1/2}}\right]. \tag{7-6.24}$$

To find the corresponding electric field in the Σ frame, we transform E_x', R', λ', and L' by using Eqs. (7-1.23), (7-1.20), (7-1.29), and (7-1.19) (observe that λ' transforms like ρ', R' transforms like y', and L' transforms like x'). Selecting $t = 0$ for the time of observation in Σ (we can choose t at will because the charge is time-independent in Σ') and noting that $J'_x = 0$ because the charge is stationary in Σ', we obtain from Eq. (7-6.24) after elementary simplifications

$$E_x = \frac{\lambda(1-v^2/c^2)}{4\pi\varepsilon_0 R}\left[\frac{1}{(L_1^2/R^2+1-v^2/c^2)^{1/2}} - \frac{1}{(L_2^2/R^2+1-v^2/c^2)^{1/2}}\right]. \tag{7-6.25}$$

Note that Eq. (7-6.25) is exactly the same as Eq. (4-3.13) that we obtained by using the retarded field calculations.

By Eq. (7-6.23), the y component of the electric field in the Σ' frame, where the charge is stationary, is

$$E_y' = -\frac{\lambda'}{4\pi\varepsilon_0 R'^2}\left[\frac{L'_1}{(L'^2_1/R'^2+1)^{1/2}} - \frac{L'_2}{(L'^2_2/R'^2+1)^{1/2}}\right]. \quad (7\text{-}6.26)$$

Using Eqs. (7-1.24), (7-1.20), (7-1.29), and (7-1.19) for transforming E_y', R', λ', and L' and taking into account that there is no magnetic field in Σ' (because the charge is at rest there) we obtain from Eq. (7-6.26) after elementary simplifications

$$E_y = \frac{\lambda}{4\pi\varepsilon_0 R^2}\left[\frac{L_2}{(L_2^2/R^2+1-v^2/c^2)^{1/2}} - \frac{L_1}{(L_1^2/R^2+1-v^2/c^2)^{1/2}}\right], \quad (7\text{-}6.27)$$

which also is exactly the same as Eq. (4-3.22) obtained from classical calculations.

Example 7-6.4 The scalar potential of a stationary charge distribution can be found from the well-known equation

$$\varphi = \frac{1}{4\pi\varepsilon_0}\int\frac{\rho}{r}dV. \quad (7\text{-}6.28)$$

Convert Eq. (7-6.28) into the scalar potential produced by a charge distribution moving with constant velocity $\mathbf{v} = v\,\mathbf{i}$.

Consider a charge distribution ρ' at rest in a reference frame Σ' which moves with respect to the laboratory with uniform velocity $\mathbf{v} = v\mathbf{i}$. The electric potential φ' produced by ρ' in this reference frame is given by Eq. (7-6.28) with φ, ρ, r, and dV replaced by the corresponding primed quantities, that is

$$\varphi' = \frac{1}{4\pi\varepsilon_0}\int\frac{\rho'}{r'}dV'. \quad (7\text{-}6.29)$$

To find the potential in the laboratory, we transform the primed quantities in Eq. (7-6.29) into the corresponding unprimed quantities. Setting $t = 0$ and using Eqs. (7-1.19)-(7-1.21), (7-1.11), and (7-1.15) with $J_x' = 0$ and $A_x' = 0$ (because there is no current and no magnetic field in Σ'), we obtain

$$\frac{\varphi}{\gamma} = \frac{1}{4\pi\varepsilon_0}\int \frac{\rho/\gamma}{(\gamma^2x^2+y^2+z^2)^{1/2}}\,d(\gamma x)dydz\,, \qquad (7\text{-}6.30)$$

or, simplifying and using Eqs. (5-1.8) and (5-1.9),

$$\varphi = \frac{1}{4\pi\varepsilon_0}\int \frac{\rho}{r[1-(v^2/c^2)\sin^2\theta]^{1/2}}\,dV\,, \qquad (7\text{-}6.31)$$

which, except for notation, is the same as Eq.(5-2.5) that was obtained from a retarded potential integral.

Example 7-6.5 The scalar potential of a stationary charge distribution whose charge density is constant throughout the volume occupied by the distribution can be found from the equation[14]

$$\varphi = -\frac{\rho}{8\pi\varepsilon_0}\oint \frac{\mathbf{r}}{r}\cdot d\mathbf{S}_{out}\,, \qquad (7\text{-}6.32)$$

where $d\mathbf{S}_{out}$ is a surface element vector directed from the charge distribution into the surrounding space. Convert Eq. (7-6.32) into the scalar potential produced by a charge distribution moving with constant velocity $\mathbf{v} = v\mathbf{i}$.

Consider a charge distribution ρ' at rest in a reference frame Σ' which moves with respect to the laboratory (reference frame Σ) with uniform velocity $\mathbf{v} = v\mathbf{i}$. The electric potential φ' produced by ρ' in this reference frame is given by Eq. (7-6.32) with φ, ρ, r, $\mathbf{r}$, and $d\mathbf{S}$ replaced by the corresponding primed quantities:

$$\varphi' = -\frac{\rho'}{8\pi\varepsilon_0}\oint \frac{\mathbf{r}'}{r'}\cdot d\mathbf{S}'_{out}\,. \qquad (7\text{-}6.33)$$

To find the potential in the laboratory, we transform the primed quantities in Eq. (7-6.33) into the equivalent expressions in terms of unprimed quantities. First, however, we expand the dot product in Eq. (7-6.33), obtaining

$$\varphi' = -\frac{\rho'}{8\pi\varepsilon_0}\oint\frac{(x'dy'dz' + y'dz'dx' + z'dx'dy')_{out}}{r'}. \quad (7\text{-}6.34)$$

Now, setting $t = 0$ and using Eqs. (7-1.19)-(7-1.21), (7-1.11), and (7-1.15) with $J_x' = 0$ and $A_x' = 0$ (because there is no current and no magnetic field in Σ'), we have

$$\frac{\varphi}{\gamma} = -\frac{\rho/\gamma}{8\pi\varepsilon_0}\oint\frac{(\gamma x dy dz + y dz \gamma dx + z\gamma dx dy)_{out}}{(\gamma x^2 + y^2 + z^2)^{1/2}}, \quad (7\text{-}6.35)$$

or, simplifying, using Eqs. (5-1.8) and (5-1.9), and restoring the vector notation,

$$\varphi = -\frac{\rho}{8\pi\varepsilon_0}\oint\frac{\mathbf{r}\cdot d\mathbf{S}_{out}}{r[1-(v^2/c^2)\sin^2\theta]^{1/2}}, \quad (7\text{-}6.36)$$

which, except for notation, is the same as Eq. (5-2.11) obtained from a retarded potential integral. ▲

References and Remarks for Chapter 7

1. The history of the relativistic transformation equations for coordinates and time is quite interesting. The transformation of coordinates that we now call "Lorentz transformations" was apparently invented by J. J. Thomson who used it in the article "On the magnetic effects produced by motion in the electric field," Philos. Mag. **46**, 528-545 (1889). H. A. Lorentz postulated the transformation equations (7-1.1)-(7-1.3) with $t = 0$ in his book *Versuch einer Theorie der Elektrischen und Optischen Erscheinungen in Bewegten Körper* (E. J. Brill, Leiden, 1895). J. Larmor used the complete set of transformations of coordinates and time in the paper "On a dynamical theory of the electric and luminiferous medium," Philos. Trans. R. Soc. London **190A**, 205

(1897). J. Larmor again used the complete set of transformations of coordinates and time in his book *Aether and Matter* (Cambridge U. P., Cambridge, 1900), where he for the first time applied them for transforming Maxwell's equations. H. A. Lorentz also used these equations for transforming Maxwell's equations in his article "Electromagnetic Phenomena in a System Moving with any Velocity less than Light," Proc. Acad. Sci. Amsterdam **6**, 809 (1904). The name "Lorentz transformations" first appeared in H. Poincaré, "Sur la dynamique de l'électron," C. R. Acad. Sci. **140**, 1504-1508 (1905). It should be noted that similar transformations appeared in W. Voigt, "Über das Doppler'sche Princip," Göttinger Gesellschaft der Wissenschaften - Nachrichten 41-51 (1887).

2. The transformation equations for electric and magnetic fields (in a different notation) first appeared in H. A. Lorentz, "Simplified Theory of Electrical and Optical Phenomena in Moving Systems," Koninkl. Acad. Wetenschap. Proc. (English ed.) **1**, 427-443 (1899). J. Larmor obtained these equations (likewise in a different notation) in his *Aether and Matter* (Cambridge U. P., Cambridge, 1900). H. A. Lorentz published the equations again in his article "Electromagnetic Phenomena in a System Moving with any Velocity less than Light," Proc. Acad. Sci. Amsterdam **6**, 809-834 (1904). The transformation equation for the charge density used by Lorentz was incomplete (the convection current density J_x, or ρv_x, was absent). H. Poincaré gave the complete version of this equation in his article "Sur la dynamique de l'électron," C. R. Acad. Sci. **140**, 1504-1508 (1905). A. Einstein gave the complete version of this equation in his article "Zur Elektrodynamik bewegter Körper," Ann. Phys. **17**, 891-921 (1905).

3. See H. Poincaré, "Sur la dynamique de l'électron," Ren. Circ. Mat. Palermo **21**, 129-175 (1906).

4. Relativistic velocity transformation equations, also known as Einstein's velocity addition law, were first derived by A. Einstein in "Zur Elektrodynamik bewegter Körper," Ann. Phys. **17**, 891-921 (1905) and by H. Poincaré in "Sur la dynamique de l'électron," Ren. Circ. Mat. Palermo **21**, 129-175 (1906).

5. Relativistic transformation equations for acceleration were first derived by H. Poincaré. See Ref. 3.

6. The invariance of the Cartesian components of Maxwell's equations under what we now call "Lorentz transformations" was first demonstrated by J. Larmor in *Aether and Matter* (Cambridge U. P., Cambridge, 1900). H. A. Lorentz showed the invariance of the Cartesian components of Maxwell's equations under these transformations in the article "Electromagnetic Phenomena in a System Moving with any Velocity less than Light," Proc. Acad. Sci. Amsterdam **6**, 809-834 (1904). A. Einstein showed the invariance in the article "Zur Elektrodynamik bewegter Körper," Ann. Phys. **17**, 891-921 (1905). H. Poincaré showed the invariance (using electromagnetic potentials) in the article "Sur la dynamique de l'électron," Ren. Circ. Mat. Palermo **21**, 129-175 (1906).

7. The noninvariance of Maxwell's Eqs. (2-1.3) and (2-1.4) under relativistic transformations is ignored in conventional presentations of relativity theory, where the authors erroneously assert that "Maxwell's equations are invariant (or covariant) under Lorentz transformations." In fact, Lorentz, Poincaré, and Einstein used Maxwell's equations in their scalar form and therefore only showed that the Cartesian components of Maxwell's equations were invariant under the transformations.

8. It should be noted, however, that, just as in the case of Maxwell's equations [see Section (7-4)], the transformations do not work if applied directly to the vector equation (7-5.21) for $\mathbf{E}'$.

9. H. A. Lorentz, *Versuch einer Theorie der elektrischen und optischen Erscheinungen in bewegten Körpern* (Brill, Leiden, 1895). For a discussion see K. F. Schaffner, "The Lorentz electron theory of relativity," Am. J. Phys. **37**, 498-513 (1969).

10. See also Oleg D. Jefimenko, "Retardation and relativity: new integrals for electric and magnetic potentials of time-independent charge distributions moving with constant velocity," Eur. J. Phys. **17**, 258-264 (1996).

11. See Oleg D. Jefimenko, *Electricity and Magnetism*, 2nd. ed., (Electret Scientific, Star City, 1989) pp. 101-103.

12. See, for example, Ref. 11, pp. 92-93.

13. See, for example, Ref. 11, pp. 98-99.

14. See, for example, Ref. 11, p. 141.

8

FROM RELATIVISTIC ELECTROMAGNETISM TO RELATIVISTIC MECHANICS

Electric and magnetic fields are force fields. They exert forces on charged bodies and affect the state of motion of these bodies. The study of the motion of bodies under the action of different forces is the domain of mechanics. However, classical mechanics was developed much earlier than electromagnetic theory and before the advent of relativistic electrodynamics. It is clear therefore that classical mechanics needs to be reformulated to make it compatible with relativistic electrodynamics. The mechanics thus reformulated is called *relativistic mechanics*. Its fundamentals are presented in this chapter on the basis of already developed relations of relativistic electrodynamics.

8-1. Transformation of the Lorentz Force

In Chapter 7 we derived relativistic transformation equations for electric and magnetic fields. Electric and magnetic fields are force fields. We may expect, therefore, that our transformation equations for electric and magnetic fields could be converted into force transformation equations. To explore this possibility we shall proceed as follows.[1,2]

The force experienced by a point charge q moving with velocity $\mathbf{u}$ in the presence of an electric field $\mathbf{E}$ and a magnetic flux density field $\mathbf{B}$ is given by the Lorentz force law[3]

$$\mathbf{F} = q(\mathbf{E} + \mathbf{u} \times \mathbf{B}). \tag{8-1.1}$$

This law does not depend on the inertial reference frame in which q, $\mathbf{u}$, $\mathbf{E}$, and $\mathbf{B}$ are measured. Therefore in an inertial reference frame Σ' moving with velocity v relative to the laboratory (reference frame Σ) in the direction of their common x axis, Lorentz force law can be written as

$$\mathbf{F}' = q(\mathbf{E}' + \mathbf{u}' \times \mathbf{B}'), \tag{8-1.2}$$

where the primes are used to indicate quantities measured in the moving reference frame (there is no prime on q because the charge does not depend on the velocity with which it moves). All we need to do to obtain an equation transforming $\mathbf{F}'$ to $\mathbf{F}$ is to express $\mathbf{E}$, $\mathbf{u}$, and $\mathbf{B}$ in Eq. (8-1.1) in terms of primed quantities and to group the latter together in the form of Eq. (8-1.2). However, when dealing with relativistic transformation, it is usually much simpler to write the transformation equations in terms of the Cartesian components of the vectors involved rather than in terms of the vectors themselves. In terms of the components, Eqs. (8-1.1) and (8-1.2) are

$$F_x = q(E_x + u_y B_z - u_z B_y), \tag{8-1.3}$$

$$F_y = q(E_y + u_z B_x - u_x B_z), \tag{8-1.4}$$

$$F_z = q(E_z + u_x B_y - u_y B_x); \tag{8-1.5}$$

and

$$F_x' = q(E_x' + u_y' B_z' - u_z' B_y'), \tag{8-1.6}$$

$$F_y' = q(E_y' + u_z' B_x' - u_x' B_z'), \tag{8-1.7}$$

$$F_z' = q(E_z' + u_x' B_y' - u_y' B_x'). \tag{8-1.8}$$

Transformation equation for the x component of **F.** Substituting Eqs. (7-1.5), (7-2.6), (7-1.10), (7-2.7), and (7-1.9) into Eq. (8-1.3) and cancelling gamma, we have

$$F_x = q\left[E_x' + \frac{u_y'}{1+vu_x'/c^2}\left(B_z' + \frac{vE_y'}{c^2}\right) - \frac{u_z'}{1+vu_x'/c^2}\left(B_y' - \frac{vE_z'}{c^2}\right)\right]. \quad (8\text{-}1.9)$$

Adding and subtracting

$$\frac{vu_y'u_z'B_x'}{c^2(1+vu_x'/c^2)},$$

we obtain

$$\begin{aligned}
F_x &= q\left[E_x' + \frac{u_y'}{1+vu_x'/c^2}\left(B_z' + \frac{vE_y'}{c^2}\right) - \frac{u_z'}{1+vu_x'/c^2}\left(B_y' - \frac{vE_z'}{c^2}\right)\right. \\
&\quad \left. + \frac{vu_y'u_z'B_x'}{c^2(1+vu_x'/c^2)} - \frac{vu_y'u_z'B_x'}{c^2(1+vu_x'/c^2)}\right] \\
&= q\left[E_x' + \frac{u_y'}{1+vu_x'/c^2}\left(B_z' + \frac{vE_y'}{c^2} + \frac{vu_z'B_x'}{c^2}\right)\right. \qquad (8\text{-}1.10) \\
&\quad \left. - \frac{u_z'}{1+vu_x'/c^2}\left(B_y' - \frac{vE_z'}{c^2} + \frac{vu_y'B_x'}{c^2}\right)\right] \\
&= q\left[E_x' + \frac{vu_y'}{c^2(1+vu_x'/c^2)}\left(\frac{c^2}{v}B_z' + E_y' + u_z'B_x'\right)\right. \\
&\quad \left. - \frac{vu_z'}{c^2(1+vu_x'/c^2)}\left(\frac{c^2}{v}B_y' - E_z' + u_y'B_x'\right)\right].
\end{aligned}$$

Adding and subtracting $u_x'B_z'$ inside the parentheses of the first term and $u_x'B_y'$ inside the parentheses of the second term of the last expression, we then have

$$F_x = q\left[E_x' + \frac{vu_y'}{c^2(1+vu_x'/c^2)}\left(\frac{c^2}{v}B_z' + u_x'B_z' - u_x'B_z' + E_y' + u_z'B_x'\right)\right.$$
$$\left. - \frac{vu_z'}{c^2(1+vu_x'/c^2)}\left(\frac{c^2}{v}B_y' + u_x'B_y' - u_x'B_y' - E_z' + u_y'B_x'\right)\right]. \tag{8-1.11}$$

Simplifying Eq. (8-1.11), we obtain

$$F_x = q\left[E_x' + \frac{vu_y'}{c^2(1+vu_x'/c^2)}\left(\frac{c^2(1+vu_x'/c^2)}{v}B_z' - u_x'B_z' + E_y' + u_z'B_x'\right)\right.$$
$$\left. - \frac{vu_z'}{c^2(1+vu_x'/c^2)}\left(\frac{c^2(1+vu_x'/c^2)}{v}B_y' - u_x'B_y' - E_z' + u_y'B_x'\right)\right], \tag{8-1.12}$$

or

$$F_x = q\left[E_x' + u_y'B_z' - u_z'B_y' + \frac{vu_y'}{c^2(1+vu_x'/c^2)}\left(E_y' + u_z'B_x' - u_x'B_z'\right)\right.$$
$$\left. + \frac{vu_z'}{c^2(1+vu_x'/c^2)}\left(E_z' + u_x'B_y' - u_y'B_x'\right)\right]. \tag{8-1.13}$$

Comparing Eq. (8-1.13) with Eqs. (8-1.6), (8-1.7), and (8-1.8), we recognize that Eq. (8-1.13) can be written as

$$F_x = F_x' + \frac{vu_y'}{c^2(1+vu_x'/c^2)}F_y' + \frac{vu_z'}{c^2(1+vu_x'/c^2)}F_z', \tag{8-1.14}$$

which is the transformation equation for obtaining the x component of the force measured in the laboratory system from the x, y, and z components of the force measured in the moving system.

***Transformation equation for the y component of* F.** Substituting Eqs. (7-1.6), (7-2.7), (7-1.8), (7-2.5), and (7-1.10) into Eq. (8-1.4), we have

$$F_y = q\left[\gamma(E_y' + vB_z') + \frac{u_z'}{\gamma(1 + vu_x'/c^2)}B_x' - \frac{u_x' + v}{1 + vu_x'/c^2}\gamma\left(B_z' + \frac{v}{c^2}E_y'\right)\right]. \tag{8-1.15}$$

Factoring out

$$\frac{\gamma}{1 + vu_x'/c^2},$$

simplifying, and rearranging, we obtain

$$\begin{aligned} F_y &= \frac{q\gamma}{1 + vu_x'/c^2}\left[(E_y' + vB_z')\left(1 + \frac{vu_x'}{c^2}\right)\right. \\ &\quad \left. + u_z'\left(1 - \frac{v^2}{c^2}\right)B_x' - (u_x' + v)\left(B_z' + \frac{v}{c^2}E_y'\right)\right] \\ &= \frac{q\gamma}{1 + vu_x'/c^2}\left[\left(1 - \frac{v^2}{c^2}\right)E_y' - u_x'\left(1 - \frac{v^2}{c^2}\right)B_z' + u_z'\left(1 - \frac{v^2}{c^2}\right)B_x'\right] \\ &= \frac{q}{\gamma(1 + vu_x'/c^2)}(E_y' + u_z'B_x' - u_x'B_z'), \end{aligned} \tag{8-1.16}$$

or, with Eq. (8-1.7),

$$F_y = \frac{1}{\gamma(1 + vu_x'/c^2)}F_y', \tag{8-1.17}$$

which is the transformation equation for obtaining the y component of the force measured in the laboratory system from the y component of the force measured in the moving system.

***Transformation equation for the z component of* F.** Substituting Eqs. (7-1.7), (7-2.5), (7-1.9), (7-2.6), and (7-1.8) into Eq. (8-1.5) and proceeding as we did for deriving Eq. (8-1.17), we get

$$F_z = \frac{1}{\gamma(1 + vu_x'/c^2)}F_z', \tag{8-1.18}$$

which is the transformation equation for obtaining the z

component of the force measured in the laboratory system from the z component of the force measured in the moving system.

***Inverse transformation equations for* F.** The transformation equations that we have obtained are for transforming forces from the moving (primed) reference frame to the laboratory (stationary) reference frame. The inverse transformations can be derived in the same manner. However, as usual, the inverse transformations can be obtained without additional derivations by simply switching primes from the primed to the unprimed quantities and reversing the sign in front of v. The result is

$$F_x' = F_x - \frac{vu_y}{c^2(1 - vu_x/c^2)}F_y - \frac{vu_z}{c^2(1 - vu_x/c^2)}F_z, \tag{8-1.19}$$

$$F_y' = \frac{1}{\gamma(1 - vu_x/c^2)}F_y, \tag{8-1.20}$$

and

$$F_z' = \frac{1}{\gamma(1 - vu_x/c^2)}F_z. \tag{8-1.21}$$

8-2. Transformation of Electromagnetic Energy and Momentum

If an electric point charge q is located in a region of an electric field where the electric potential is φ, the energy of the interaction between q and the electric field (the potential energy of q) is[4]

$$U_e = q\varphi. \tag{8-2.1}$$

Likewise, if a point charge q moves with velocity $\mathbf{u}$ in a region of a magnetic field where the magnetic vector potential is $\mathbf{A}$, the energy of the interaction between q and the magnetic field is[5]

$$U_m = q\mathbf{u} \cdot \mathbf{A}. \tag{8-2.2}$$

Also, if a point charge q is located in a time-independent magnetic field in a region where the magnetic vector potential is $\mathbf{A}$, there exists an electromagnetic momentum associated with q and with the magnetic field

$$\mathbf{G} = q\mathbf{A}. \tag{8-2.3}$$

Equation (8-2.3) can be obtained as follows. The electromagnetic momentum contained in an electromagnetic field is[6]

$$\mathbf{G} = \varepsilon_0\mu_0\int \mathbf{E}\times\mathbf{H}\,dV, \tag{8-2.4}$$

where $\mathbf{E}$ is the electric field and $\mathbf{H}$ is the magnetic field, and the integration is extended over the region where the two fields are present. Since in a vacuum, by Eqs. (2-1.5) and (2-1.6), $\mu_0\mathbf{H} = \mathbf{B}$ and $\varepsilon_0\mathbf{E} = \mathbf{D}$, and since, by Eq. (2-4.1), $\mathbf{B} = \nabla\times\mathbf{A}$, where $\mathbf{A}$ is the vector potential of $\mathbf{B}$, we can write Eq. (8-2.4) as

$$\mathbf{G} = \int \mathbf{D}\times(\nabla\times\mathbf{A})\,dV. \tag{8-2.5}$$

Consider now the case when $\mathbf{D}$ is due to an electric charge distribution ρ moving with velocity $\mathbf{u}$ through a time-independent magnetic field. Using vector identity (V-22), we obtain

$$\oint(\mathbf{D}\cdot\mathbf{A})d\mathbf{S} - \oint\mathbf{A}(\mathbf{D}\cdot d\mathbf{S}) - \oint\mathbf{D}(\mathbf{A}\cdot d\mathbf{S}) \tag{8-2.6}$$
$$= \int[\mathbf{D}\times(\nabla\times\mathbf{A}) + \mathbf{A}\times(\nabla\times\mathbf{D}) - \mathbf{D}(\nabla\cdot\mathbf{A}) - \mathbf{A}(\nabla\cdot\mathbf{D})]\,dV,$$

where the integration is over all space. Now, since $\mathbf{D}$ and $\mathbf{A}$ vanish at infinity at least as $1/r^2$, the surface integrals vanish. Also, by Eq. (2-1.1), $\nabla\cdot\mathbf{D} = \rho$, and since the magnetic field is time independent, by Eq. (2-1.3), $\nabla\times\mathbf{D} = 0$, and by Eq. (2-4.9), $\nabla\cdot\mathbf{A} = 0$. Therefore Eq. (8-2.6) reduces to

$$\int\mathbf{D}\times(\nabla\times\mathbf{A})\,dV = \int\mathbf{A}\rho\,dV. \tag{8-2.7}$$

If the charge under consideration is a point charge, **A** can be considered constant throughout the volume occupied by the charge and can be factored out from the last integral in Eq. (8-2-7). Since $\int \rho dV = q$, we then obtain Eq. (8-2.3) from Eqs. (8-2.7) and (8-2.5).

Let us now assume that a point charge q is at rest in a reference frame Σ' which moves with velocity $\mathbf{v} = v\mathbf{i}$ relative to the laboratory (reference frame Σ). Let the charge be located in an external electric field whose potential, as measured in Σ', is φ'. The energy of the interaction between q and the electric field in Σ' is then

$$U_e' = q\varphi' . \tag{8-2.8}$$

Let us now express U_e' in terms of the quantities measured in the laboratory. Using Eq. (7-1.33) for transforming φ', we have

$$U_e' = q\gamma(\varphi - vA_x) \tag{8-2.9}$$

or

$$U_e' = \gamma(q\varphi - qvA_x). \tag{8-2.10}$$

However, by Eq. (8-2.1), $q\varphi$ is the energy of the interaction between q and the electric field as measured in the laboratory, and by Eq. (8-2.2), qvA_x is the energy of the interaction between q and the magnetic field as measured in the laboratory. Hence the transformation equation for the electric interaction energy is

$$U_e' = \gamma(U_e - U_m). \tag{8-2.11}$$

As usual, the inverse transformation equation is

$$U_e = \gamma(U_e' + U_m') \tag{8-2.12}$$

[the "+" sign follows from Eq. (8-2.10), where there is a "−" in front of qv].

Observe that instead of interpreting the term qvA_x in Eq. (8-2.10) as the magnetic interaction energy, we can interpret it, according to Eq. (8-2.3), as the product of v and the x component of the electromagnetic momentum G_x. Therefore we can also write Eq. (8-2.11) as

$$U_e' = \gamma(U_e - vG_x), \tag{8-2.13}$$

and Eq. (8-2.12) as

$$U_e = \gamma(U_e' + vG_x'). \tag{8-2.14}$$

Let us now obtain transformation equations for the electromagnetic momentum **G**. Writing Eq. (8-2.3) in terms of Cartesian components and using Eqs. (7-1.34)-(7-1.36), we can express the electromagnetic momentum **G**′ measured in Σ' in terms of the electric and magnetic potentials measured in the laboratory as

$$G_x' = q\gamma[A_x - (v/c^2)\varphi], \tag{8-2.15}$$

$$G_y' = qA_y, \tag{8-2.16}$$

$$G_z' = qA_z. \tag{8-2.17}$$

However, qA_x is the x component of the electromagnetic momentum associated with q as measured in the laboratory, and $q\varphi$ is the electric interaction energy (the potential energy of q) as measured in the laboratory. Hence, by Eqs. (8-2.15)-(8-2.17) we have for the transformation of electromagnetic momentum

$$G_x' = \gamma[G_x - (v/c^2)U_e], \tag{8-2.18}$$

$$G_y' = G_y, \tag{8-2.19}$$

$$G_z' = G_z. \tag{8-2.20}$$

The inverse transformation equations are then

$$G_x = \gamma[G_x' + (v/c^2)U_e'], \tag{8-2.21}$$

$$G_y = G_y', \tag{8-2.22}$$

$$G_z = G_z'. \tag{8-2.23}$$

8-3. Relativistic Expression for Mechanical Momentum

Let a charged particle of mass m move with velocity $\mathbf{u} = u_x\mathbf{i}$ in the laboratory reference frame Σ. Observed in a reference frame Σ' which moves with velocity $\mathbf{v} = v\mathbf{i} = u_x\mathbf{i}$ relative to the laboratory, the particle is at rest. Let there be an electric field in Σ' acting on the particle with a force $\mathbf{F}'$. Since the particle is at rest in Σ', it obeys the well-known laws of classical mechanics there. In particular, it experiences an acceleration under the action of $\mathbf{F}'$ according to Newton's second law, so that, considering the x component of the force, we have

$$F_x' = ma_x' \tag{8-3.1}$$

where a_x' is the acceleration of the particle in Σ' (note that although the particle is at rest in Σ', so that $u_x' = 0$, $a_x' \neq 0$ as long as a force acts upon the particle).

Let us now convert Eq. (8-3.1) to the laboratory reference frame. Taking into account that $u_x' = u_y' = u_z' = 0$ and that $v = u_z$, and replacing in Eq. (8-3.1) F_x' by F_x and a_x' by a_x with the help of Eqs. (8-1.14) and (7-2.14), we have

$$F_x = \gamma^3 m a_x. \tag{8-3.2}$$

Consider now the relation

$$\begin{aligned}\frac{d}{dt}\left[\frac{u_x}{(1-u_x^2/c^2)^{1/2}}\right] &= \frac{(1-u_x^2/c^2)^{1/2}du_x/dt + [u_x^2/c^2(1-u_x^2/c^2)^{1/2}]du_x/dt}{1-u_x^2/c^2} \\ &= \frac{1}{(1-u_x^2/c^2)^{3/2}}\frac{du_x}{dt} = \frac{1}{(1-u_x^2/c^2)^{3/2}}a_x.\end{aligned} \tag{8-3.3}$$

Since by supposition Σ' moves with velocity $\mathbf{v} = u_x\mathbf{i}$, so that $u_x = v$, the fraction in the last term is the same as γ^3 so that we can write

$$\frac{d}{dt}\left[\frac{u_x}{(1-u_x^2/c^2)^{1/2}}\right] = \gamma^3 a_x . \tag{8-3.4}$$

Combining Eqs. (8-3.2) and (8-3.4), we obtain

$$F_x = \frac{d}{dt}\left[\frac{mu_x}{(1-u_x^2/c^2)^{1/2}}\right]. \tag{8-3.5}$$

But, by Newton's second law, the force acting on a body is equal to the rate of change of the mechanical momentum of the body. Therefore the x component of the mechanical momentum of the particle under consideration is not $p_x = mu_x$, as it is defined in classical mechanics, but

$$p_x = \frac{mu_x}{(1-u_x^2/c^2)^{1/2}} . \tag{8-3.6}$$

For the y component of the force acting on the particle in Σ' we have

$$F_y' = ma_y' . \tag{8-3.7}$$

Let us convert Eq. (8-3.7) to the laboratory reference frame. Taking into account that $u_x' = u_y' = u_z' = 0$ and that $v = u_z$, and replacing in Eq. (8-3.7) F_y' by F_y and a_y' by a_y with the help of Eqs. (8-1.17) and (7-2.15), we have

$$\gamma F_y = \gamma^2 ma_y \tag{8-3.8}$$

or

$$F_y = \gamma ma_y . \tag{8-3.9}$$

Consider now the relation

$$\frac{d}{dt}\left[\frac{u_y}{(1-u_x^2/c^2)^{1/2}}\right] = \frac{(1-u_x^2/c^2)^{1/2}du_y/dt + [u_y u_x/c^2(1-u_x^2/c^2)^{1/2}]du_x/dt}{1-u_x^2/c^2}$$

$$= \frac{1}{(1-u_x^2/c^2)^{1/2}}du_y/dt = \frac{1}{(1-u_x^2/c^2)^{1/2}}a_y \tag{8-3.10}$$

(in obtaining this relation we took into account that $u_y = 0$, because by supposition $\mathbf{u} = v\mathbf{i}$, so that only the x component of $\mathbf{u}$ is different from zero). Since $u_x = v$, the fraction in the last term of Eq. (8-3.10) is the same as γ, so that we can write

$$\frac{d}{dt}\left[\frac{u_y}{(1-u_x^2/c^2)^{1/2}}\right] = \gamma a_y. \tag{8-3.11}$$

Combining Eqs. (8-3.9) and (8-3.11), we obtain

$$F_y = \frac{d}{dt}\left[\frac{mu_y}{(1-u_x^2/c^2)^{1/2}}\right]. \tag{8-3.12}$$

Therefore the y component of the mechanical momentum of the particle under consideration is not $p_y = mu_y$, as it is defined in classical mechanics, but

$$p_y = \frac{mu_y}{(1-u_x^2/c^2)^{1/2}}. \tag{8-3.13}$$

By the same procedure we find that the z component of the mechanical momentum of the particle is

$$p_z = \frac{mu_z}{(1-u_x^2/c^2)^{1/2}}. \tag{8-3.14}$$

Combining Eqs. (8-3.6), (8-3.13), and (8-3.14) into a single vector equation, and remembering that by supposition $\mathbf{u} = u_x\mathbf{i}$ and $\mathbf{v} = u_x\mathbf{i}$, so that $\mathbf{u} = \mathbf{v}$, we obtain for the relativistic momentum of a particle of mass m moving with velocity $\mathbf{u}$[7]

$$\mathbf{p} = \frac{m\mathbf{u}}{(1 - u^2/c^2)^{1/2}}. \tag{8-3.15}$$

Observe that if a particle moves with a velocity much smaller than c, its relativistic momentum reduces to the classical mechanical momentum

$$\mathbf{p}_{u \ll c} = m\mathbf{u}. \tag{8-3.16}$$

8-4. Relativistic Mass, Longitudinal Mass, and Transverse Mass

We can write Eq. (8-3.15) in the classical form by introducing the concept of a velocity-dependent *relativistic mass*, defined as

$$m_r = \frac{m}{(1 - u^2/c^2)^{1/2}}, \tag{8-4.1}$$

where m is the ordinary mass measured when the body under consideration is at rest, sometimes called the *proper mass*, or the *rest mass*. In terms of relativistic mass, the mechanical momentum given by Eq. (8-3.15) becomes

$$\mathbf{p} = m_r\mathbf{u}. \tag{8-4.2}$$

The utility of the concept of relativistic mass is highly questionable and we shall not use the concepts or expressions "relativistic mass," "proper mass," or "rest mass" in this book.[8]

There are, however, two other "masses" in relativity theory, which occasionally have useful applications (see Chapter 10). Their meaning is explained below.

The primary significance of Eq. (8-3.15) is that with the help of this equation it becomes possible to determine the acceleration, velocity, and trajectory of particles moving under the influence of external forces with speeds close to c. From Newton's second law and Eq. (8-3.15), we have

$$\mathbf{F} = \frac{d\mathbf{p}}{dt} = \frac{d}{dt}\left[\frac{m\mathbf{u}}{(1-u^2/c^2)^{1/2}}\right]. \tag{8-4.3}$$

Differentiating, we have

$$\mathbf{F} = \frac{m}{(1-u^2/c^2)^{1/2}}\frac{d\mathbf{u}}{dt} + \frac{mu\mathbf{u}}{c^2(1-u^2/c^2)^{3/2}}\frac{du}{dt}. \tag{8-4.4}$$

Using Eq. (8-4.3), we can also write

$$\begin{aligned}\left(\frac{\mathbf{F}\cdot\mathbf{u}}{c^2}\right)\mathbf{u} &= \left\{\frac{d}{dt}\left[\frac{m\mathbf{u}}{(1-u^2/c^2)^{1/2}}\right]\cdot\frac{\mathbf{u}}{c^2}\right\}\mathbf{u} = \left\{\frac{d}{dt}\left[\frac{mu}{(1-u^2/c^2)^{1/2}}\right]\frac{u}{c^2}\right\}\mathbf{u}\\ &= \left[\frac{mu\mathbf{u}}{c^2(1-u^2/c^2)^{1/2}}\right]\frac{du}{dt} + \left[\frac{mu^2\mathbf{u}u/c^2}{c^2(1-u^2/c^2)^{3/2}}\right]\frac{du}{dt} \qquad (8\text{-}4.5)\\ &= \frac{mu\mathbf{u}}{c^2(1-u^2/c^2)^{1/2}}\left[\frac{1-u^2/c^2+u^2/c^2}{1-u^2/c^2}\right]\frac{du}{dt} = \frac{mu\mathbf{u}}{c^2(1-u^2/c^2)^{3/2}}\frac{du}{dt}\end{aligned}$$

or

$$\frac{mu\mathbf{u}}{c^2(1-u^2/c^2)^{3/2}}\frac{du}{dt} = \left(\frac{\mathbf{F}\cdot\mathbf{u}}{c^2}\right)\mathbf{u}. \tag{8-4.6}$$

Substituting Eq. (8-4.6) into Eq. (8-4.4), solving the resulting equation for $d\mathbf{u}/dt$ and replacing $d\mathbf{u}/dt$ by $\mathbf{a}$, we obtain

$$\mathbf{a} = \frac{\mathbf{F} - (\mathbf{F}\cdot\mathbf{u})\mathbf{u}/c^2}{m/(1-u^2/c^2)^{3/2}}. \tag{8-4.7}$$

Examining Eq. (8-4.7) we notice that contrary to the laws of classical mechanics, because of the presence of the vector $\mathbf{u}$ in the numerator of Eq. (8-4.7), the direction of the acceleration of a particle is, in general, not parallel to the direction of the force applied to the particle [note, however, that for particles whose velocity is much less than c, the terms with c^2 in Eq. (8-4.7) vanish, and the equation becomes the ordinary Newtonian equation of motion]. Let us now take a closer look at Eq. (8-4.7).

Let us assume that the applied force is in the direction of the velocity of the particle ("longitudinal" direction). In this case $(\mathbf{F} \cdot \mathbf{v})\mathbf{v} = Fu^2$, and Eq. (8-4.7) becomes

$$\mathbf{a}_{\parallel} = \frac{\mathbf{F}_{\parallel}(1 - u^2/c^2)}{m/(1 - u^2/c^2)^{3/2}} = \frac{\mathbf{F}_{\parallel}}{m/(1 - u^2/c^2)^{1/2}}, \tag{8-4.8}$$

so that in this case the acceleration is parallel to the force. If we now define the *longitudinal mass* as

$$m_{\parallel} = \frac{m}{(1 - u^2/c^2)^{3/2}}, \tag{8-4.9}$$

we can write Eq. (8-4.8) as

$$\mathbf{a}_{\parallel} = \frac{\mathbf{F}_{\parallel}}{m_{\parallel}}, \tag{8-4.10}$$

which, except for the subscripts "$\parallel$", looks just like the classical equation for the acceleration of a particle.

Let us now assume that the force is applied in a direction perpendicular to the velocity of the particle ("transverse" direction). In this case the last term in the numerator of Eq. (8-4.8) vanishes, and the equation becomes

$$\mathbf{a}_{\perp} = \frac{\mathbf{F}_{\perp}}{m/(1 - u^2/c^2)^{1/2}}, \tag{8-4.11}$$

so that also in this case the acceleration is parallel to the force. If we now define the *transverse mass* as

$$m_{\perp} = \frac{m}{(1 - u^2/c^2)^{1/2}}, \tag{8-4.12}$$

we can write Eq. (8-4.11) as

$$\mathbf{a}_{\perp} = \frac{\mathbf{F}_{\perp}}{m_{\perp}}, \tag{8-4.13}$$

which, except for the subscripts " ⊥ ", also looks just like the classical equation for the acceleration of a particle.[9]

In the past it was thought that the formulas for the relativistic mass, the transverse mass and the longitudinal mass indicated that the mass of a body depended on the velocity of the body. This interpretation of the formulas is now generally rejected, and the formulas are regarded merely as definitions of abbreviations that simplify the writing of certain equations but have no physical significance as such.

8-5. Transformation Equations for Mechanical Force, Energy, and Momentum

The principle of relativity demands that if a body is in equilibrium under the action of forces in a moving reference frame, it must remain in equilibrium under the action of forces in the laboratory. A charged body cannot be in a state of stable equilibrium under the action of electric forces alone (this statement is known as the "Earnshaw theorem"). Therefore mechanical forces must be present to keep the body in equilibrium. But if the transformation equations for mechanical forces are not the same as those for electric forces, then a body in a state of equilibrium in a moving reference frame will not be in equilibrium in the laboratory. Hence the transformation equations for mechanical forces are exactly the same as Eqs. (8-1.14), (8-1.17), (8-1.18), and (8-1.19)-(8-1.21), that is

$$F_x = F_x' + \frac{v u_y'}{c^2(1 + v u_x'/c^2)} F_y' + \frac{v u_z'}{c^2(1 + v u_x'/c^2)} F_z', \quad (8\text{-}5.1)$$

$$F_y = \frac{1}{\gamma(1 + v u_x'/c^2)} F_y', \quad (8\text{-}5.2)$$

$$F_z = \frac{1}{\gamma(1+vu_x'/c^2)} F_z' ; \tag{8-5.3}$$

and

$$F_x' = F_x - \frac{vu_y}{c^2(1-vu_x/c^2)} F_y - \frac{vu_z}{c^2(1-vu_x/c^2)} F_z , \tag{8-5.4}$$

$$F_y' = \frac{1}{\gamma(1-vu_x/c^2)} F_y , \tag{8-5.5}$$

$$F_z' = \frac{1}{\gamma(1-vu_x/c^2)} F_z . \tag{8-5.6}$$

By inspection we see that equations for F_x and F_x' can also be written as

$$F_x = \frac{F_x' + (v/c^2)(\mathbf{F}' \cdot \mathbf{u}')}{1+vu_x'/c^2} \tag{8-5.7}$$

and

$$F_x' = \frac{F_x - (v/c^2)(\mathbf{F} \cdot \mathbf{u})}{1-vu_x/c^2} . \tag{8-5.8}$$

Note that for $v \ll c$ these equations reduce to the ordinary equations of Newtonian mechanics (according to which a force is not affected by the motion of an inertial reference frame).

The laws of conservation of energy and momentum demand that in any interaction between bodies and electromagnetic fields both the energy and momentum must be conserved. This means that the potential energy of electric charges in an external electric field and the electromagnetic momentum associated with these charges can be converted into mechanical energy and momentum and vice versa. Therefore the transformation equations for the potential energy and momentum derived in Section 8-2, Eqs. (8-2.13), (8-2.14), (8-2.18)-(8-2.20), and (8-2.21)-(8-2.23), should be applicable to the mechanical energy and momentum of the charges, and since the charges can transfer their mechanical energy and momentum to uncharged bodies, the equations should be applicable to mechanical energy and momentum of any origin whatsoever.

If we designate the mechanical energy of a body by W and its mechanical momentum by p, then from Eqs. (8-2.13) and (8-3.14) we obtain for the mechanical energy and momentum of the body

$$W' = \gamma(W - vp_x) \tag{8-5.9}$$

and

$$W = \gamma(W' + vp_x'), \tag{8-5.10}$$

where, as usual, the primed quantities are measured in the reference frame Σ' moving with velocity v relative to the laboratory, and the unprimed quantities are measured in the laboratory.

From Eq. (8-2.18)-(8-2.20) and (8-2.21)-(8-2.23) we similarly have

$$p_x' = \gamma[p_x - (v/c^2)W], \tag{8-5.11}$$

$$p_y' = p_y, \tag{8-5.12}$$

$$p_z' = p_z. \tag{8-5.13}$$

The inverse transformation equations are

$$p_x = \gamma[p_x' + (v/c^2)W'], \tag{8-5.14}$$

$$p_y = p_y', \tag{8-5.15}$$

$$p_z = p_z'. \tag{8-5.16}$$

8-6. Transformation of Torque

Torque is defined by the equation

$$\mathbf{T} = \mathbf{r} \times \mathbf{F}, \tag{8-6.1}$$

where **F** is the force and **r** is the vector joining the pivot with the point of application of the force. In terms of Cartesian components, Eq. (8-6.1) is

$$T_x = r_y F_z - r_z F_y, \tag{8-6.2}$$

$$T_y = r_z F_x - r_x F_z, \tag{8-6.3}$$

$$T_z = r_x F_y - r_y F_x. \tag{8-6.4}$$

Let us assume that the system experiencing the torque is at rest in the reference frame Σ' which is moving with uniform velocity $\mathbf{v} = v\,\mathbf{i}$ relative to the laboratory reference frame Σ. Choosing $t = 0$ as the time of observation in Σ, transforming the components of **r** by means of Eqs. (7-1.19)-(7-1.21) (the **r** components transform like x', y', z'), and transforming the components of **F** by means of Eqs. (8-5.1)-(8-5.3) (observe that since the system under consideration is at rest in Σ', $\mathbf{u}' = 0$), we have

$$T_x = r_y'(F_z'/\gamma) - r_z'(F_y'/\gamma) = (r_y'F_z' - r_z'F_y')/\gamma, \tag{8-6.5}$$

$$T_y = r_z'F_x' - (r_x'/\gamma)(F_z'/\gamma) = r_z'F_x' - (r_x'F_z')/\gamma^2, \tag{8-6.6}$$

$$T_z = (r_x'/\gamma)(F_y'/\gamma) - r_y'F_x' = (r_x'F_y')/\gamma^2 - r_y'F_x'. \tag{8-6.7}$$

In the moving reference frame Σ' the torque components are

$$T_x' = r_y'F_z' - r_z'F_y', \tag{8-6.8}$$

$$T_y' = r_z'F_x' - r_x'F_z', \tag{8-6.9}$$

$$T_z' = r_x' F_y' - r_y' F_x' . \tag{8-6.10}$$

Comparing Eqs. (8-6.5)-(8-6.7) with Eqs. (8-6.8)-(8-6.10), and noting that $1/\gamma^2 = 1 - v^2/c^2$, we recognize that the components of **T** can be expressed in terms of the components of **T**$'$ as follows

$$T_x = T_x'/\gamma, \tag{8-6.11}$$

$$T_y = T_y' + (v^2/c^2) r_x' F_z' , \tag{8-6.12}$$

$$T_z = T_z' - (v^2/c^2) r_x' F_y' . \tag{8-6.13}$$

The inverse transformation equations (for the system at rest in Σ) are obtained by simply transposing the primes from the left to the right side of Eqs. (8-6.11)-(8-6.13).

8-7. Rest Energy, Kinetic Energy, and the Relation between Relativistic and Classical Mechanics

It is not difficult to see that there appears to be a serious problem with the transformation equations for mechanical energy and momentum derived in Section 8-5. Suppose that in the Σ' reference frame a body of mass m is at rest and no external forces act on the body. In this case W' and $\mathbf{p}'$ are zero. But then, by Eqs. (8-5.2) and (8-5.6)-(8-5.8) W and $\mathbf{p}$ are zero also in the laboratory, which is obviously wrong, since the mass *moves* relative to the laboratory (its velocity is that of the frame Σ') and thus has both kinetic energy and momentum there. Thus either our transformation equations are wrong, or the energy of a body must be different from zero even when the body is at rest and no forces act upon it.

A charged body has energy even if it is at rest and in the absence of an external electric field — its energy is due to the electric self-field of the body. Since all bodies contain electric charges in them, it is reasonable to assume that all bodies possess a certain amount of self-energy.[10] Thus there are very good reasons to assume that a body at rest possesses energy even in the absence of external forces acting upon it.

Since according to Eq. (8-3.15) the mechanical momentum of a body at rest is zero, Eq. (8-5.3) allows us to determine what the self-energy of a body should be so that our energy and momentum transformation equations would be free from contradictions. Consider a body at rest in Σ' moving with velocity $\mathbf{v} = u\mathbf{i}$ relative to the laboratory. Since $\mathbf{p}'$ is then zero and $p_x = p$, Eq. (8-5.11) can be written as

$$0 = \gamma[p - (u/c^2)W]. \tag{8-7.1}$$

Hence, solving Eq. (8-7.1) for W, we obtain the correlation between the energy and momentum of a body moving with velocity u

$$W = \frac{c^2 p}{u}, \tag{8-7.2}$$

which can also be written as

$$p = \frac{uW}{c^2}. \tag{8-7.3}$$

Let us now substitute in Eq. (8-7.2) the expression for the mechanical momentum of the body given by Eq. (8-3.15). We obtain

$$W = \frac{c^2 mu}{u(1 - u^2/c^2)^{1/2}} \tag{8-7.4}$$

so that the energy of a body moving with velocity u is

$$W = \frac{mc^2}{(1 - u^2/c^2)^{1/2}}, \tag{8-7.5}$$

from which it follows that the self-energy of a body at rest, or its "rest energy," is

$$W = mc^2. \tag{8-7.6}$$

This equation is generally known as Einstein's mass-energy equation, and is usually written as $E = mc^2$; we prefer to designate the energy by the symbol W, so as not to confuse the energy with the electric field.[11,12] Note that since the rest energy is expressed in terms of m and c, both of which are invariant under relativistic transformations, Eq. (8-7.6) holds for any inertial frame of reference.

The energy given by Eq. (8-7.5) is the *total* energy of a moving body, that is, its kinetic energy together with its rest energy. Subtracting Eq. (8-7.6) from Eq. (8-7.5), we obtain the kinetic energy

$$K = mc^2\left[\frac{1}{(1 - u^2/c^2)^{1/2}} - 1\right]. \tag{8-7.7}$$

If the velocity of the body under consideration is much smaller than c, Eq. (8-7.7) can be written as

$$K_{u \ll c} = mc^2\left[1 + u^2/2c^2 - 1\right] \tag{8-7.8}$$

or

$$K_{u \ll c} = \frac{mu^2}{2}, \tag{8-7.9}$$

which, except for the subscript, is the familiar expression for the kinetic energy of classical mechanics.

In the preceding sections of this chapter we found that equations for the relativistic force and relativistic momentum reduce to the corresponding classical expressions if the velocity of the moving reference frame or the velocity of the body under consideration is much less than c. As we have just seen, also the relativistic expression for the energy of a moving body reduces to the classical expression for the kinetic energy of the body if the velocity of the body is much less than c. It is clear therefore that relativistic mechanics has a wider range of applicability than classical mechanics, and that classical mechanics may be regarded as a subset of relativistic mechanics. On the other hand, the only presently known objects which can move with velocities comparable to that of light and can be used for experimentation are charged microscopic (atomic) particles. Therefore classical mechanics is, in general, perfectly adequate for analyzing and describing kinematic and dynamic relations involving common macroscopic bodies. However, as will be shown in Chapter 11, relativistic mechanics should be applicable also to planetary systems, including our Solar system.

▼

Example 8-7.1 Equation (8-7.4) for the mechanical energy of a moving body was derived for the laboratory reference frame. Show that it is valid for any inertial frame.

Let us transform Eq. (8-7.4) to a reference frame Σ' moving with respect to the laboratory with velocity $\mathbf{v} = v\,\mathbf{i}$. From Eqs. (8-5.2) and (8-3.16), we have

$$W' = \gamma\left[\frac{mc^2}{(1-u^2/c^2)^{1/2}} - v\frac{mu_x}{(1-u^2/c^2)^{1/2}}\right] = \gamma\frac{mc^2}{(1-u^2/c^2)^{1/2}}\left(1-\frac{vu_x}{c^2}\right). \tag{8-7.10}$$

Simplifying the last expression with the help of Eq. (7-2.24), we obtain

$$W' = \frac{mc^2}{(1-u'^2/c^2)^{1/2}}. \tag{8-7.11}$$

Thus the energy expression that we have derived for the laboratory frame is valid for all inertial frames.

Example 8-7.2 Equation (8-3.15) for the mechanical momentum of a moving body was derived for the laboratory reference frame. Show that it is valid for any inertial frame.

Consider first the x component of Eq. (8-3.15). Using Eqs. (8-5.3), (8-3.15), and (8-7.4), we can write

$$p_x' = \gamma\left[\frac{mu_x}{(1 - u^2/c^2)^{1/2}} - \left(\frac{v}{c^2}\right)\frac{mc^2}{(1 - u^2/c^2)^{1/2}}\right] = \gamma\frac{m}{(1 - u^2/c^2)^{1/2}}(u_x - v). \tag{8-7.12}$$

Multiplying and dividing the last expression by $1 - vu_x/c^2$ and using Eq. (7-2.8), we have

$$\begin{aligned} p_x' &= \gamma\frac{m}{(1 - u^2/c^2)^{1/2}}\left(\frac{u_x - v}{1 - vu_x/c^2}\right)(1 - vu_x/c^2) \\ &= \gamma\frac{mu_x'}{(1 - u^2/c^2)^{1/2}}(1 - vu_x/c^2), \end{aligned} \tag{8-7.13}$$

which, by Eq. (7-2.4), is

$$p_x' = \frac{mu_x'}{(1 - u'^2/c^2)^{1/2}}. \tag{8-7.14}$$

Consider now the y component of Eq. (8-3.15). Using Eqs. (8-5.4), (8-3.15), and (7-2.9), we can write

$$p_y' = \frac{mu_y}{(1 - u^2/c^2)^{1/2}} = \frac{mu_y'\gamma(1 - vu_x/c^2)}{(1 - u^2/c^2)^{1/2}}, \tag{8-7.15}$$

which, by Eq. (7-2.4), is

$$p_y' = \frac{mu_y'}{(1 - u'^2/c^2)^{1/2}}. \tag{8-7.16}$$

The equation for the z component of Eq. (8-3.15) is clearly similar to Eq. (8-7.15). Combining the equations for the components of p' into a single vector equation, we obtain

$$\mathbf{p}' = \frac{m\mathbf{u}'}{(1 - u'^2/c^2)^{1/2}}. \tag{8-7.17}$$

Thus the expression for mechanical momentum that we have derived for the laboratory frame is valid for all inertial frames. ▲

References and Remarks for Chapter 8

1. See also Oleg D. Jefimenko, "Derivation of relativistic force transformation equations from Lorentz force law " Am. J. Phys. **64**, 618-620 (1996).

2. Relativistic force transformation equations were first derived by M. Planck in "Das Prinzip der Relativität und die Grundgleichungen der Mechanik," Verh. D. Phys. Ges. **4**, 136-141 (1906) and in "Zur Dynamik bewegter Systeme," Berl. Ber. **13**, 542-570 (1907).

3. See, for example, Oleg D. Jefimenko, *Electricity and Magnetism*, 2nd ed., (Electret Scientific, Star City, 1989) p. 417.

4. See, for example, Ref. 3, pp. 194-195.

5. See, for example, Ref. 3, pp. 431-432. Observe that the current density associated with a moving charge is $\mathbf{J} = \rho\mathbf{v}$, and that $\int \rho dV = q$.

6. See, for example, Ref. 3, pp. 511-513.

7. This equation (in scalar form) was first derived by M. Planck. See Ref. 2.

8. For a discussion of the history and use of the concept of relativistic mass, see Carl G. Adler, "Does mass really depend on velocity, dad?" Am. J. Phys. **55**, 739-743 (1987); L. B. Okun, "The concept of mass," Phys. Today **43**, 31-36 (1989) and letters in response to this article in Phys. Today **43**, 13-15, 115-117 (1990); T. R. Sandin, "In defense of relativistic mass," Am J. Phys. **59**, 1032-1036 (1991).

9. The idea that a particle (ion) can have different effective masses for vibrations parallel and perpendicular to the velocity of translation was first suggested by H. A. Lorentz in "Simplified Theory of Electrical and Optical Phenomena in Moving Systems," Koninkl. Akad. Wetemschap. Proc. **1**, 427 (1899). The first actual (incorrect) calculation of the transverse and longitudinal masses was published by M. Abraham (who also invented the names for the two masses) in "Dynamik des Elektrons," Göttinger Nachr. 20-41 (1902). The correct expressions for the two masses were published by H. A. Lorentz in "Electromagnetic Phenomena in a System Moving with any Velocity less than Light," Proc. Acad. Sci.

Amsterdam **6**, 809-834 (1904). A. Einstein in his "Zur Elektrodynamik bewegter Körper," Ann. Phys. **17**, 891-921 (1905), referring to the idea of the two masses as "customary point of view," obtained expressions for the two masses but his expression for the transverse mass was incorrect.

10. As early as 1881, J. J. Thomson concluded that a charged body has an additional mass proportional to the electrostatic energy of the body. See J. J. Thomson, "On the electric and magnetic effects produced by the motion of electrified bodies," Philos. Mag. **11**, 229-249 (1881). Of course, it is now common knowledge that most of the self-energy of a body is not the electric but the nuclear self-energy.

11. It was Einstein who first suggested that "the mass of a body is a measure of its energy content." See A. Einstein, "Ist die Trägheit eines Körpers von seinem Energieinhalt abhängig?," Ann. Phys. **18**, 639-641 (1905). He arrived at this conclusion by considering two pulses of light emitted by a body at rest. According to Poincaré, [see H. Poincaré, "La théorie de Lorentz et la principe de réaction," in *Recuel de travaux offerts par les auteurs à H. A. Lorentz* (Nijhoff, The Hague, 1900), pp. 252-278], a pulse of light has a mass equal to E/c^2, where E is the energy of the pulse. Einstein reasoned that when light was emitted by a body, the mass of the body decreased accordingly. For the origin and history of the equation $E = mc^2$ see E. T. Whittaker, *A History of the Theories of Aether and Electricity* (Thomas Nelson, London, 1953) Vol. II, Chapt. 2 ("The Relativity Theory of Poincaré and Lorentz") pp. 51-54 and Arthur I. Miller, "*Albert Einstein's Special Theory of Relativity* (Addison-Wesley, Reading, Massachusetts, 1981) pp. 352-374 and references thereto.

12. An interesting derivation of this equation based entirely on pre-relativistic mechanics is given in J. J. Smulsky, *The Electromagnetic and Gravitational Actions* (Nauka, Novosibirsk, 1994) pp. 156-157.

9

COMMON MISCONCEPTIONS ABOUT RELATIVITY THEORY

There is a widespread belief that according to relativity theory the length of a body becomes shorter when the body moves. This is incorrect. The *length* of a body is defined as the length measured when the body is at rest relative to the observer and is an invariant quantity. There is also a widespread belief that individual relativistic transformation equations have a physical significance of their own and can be used independently one from the other. This is also wrong. Although *some* relativistic transformation equations may be used individually, in general relativistic transformation equations must be used collectively, so that *all* transformable quantities in the system under consideration are properly transformed. These and similar errors in the understanding of relativistic concepts and equations frequently result in incorrect representations of physical phenomena and in various relativistic "paradoxes" that have caused some scientists to criticize and even to reject relativity theory as such. In this chapter we shall discuss some of these errors and show the ways to avoid them.

9-1. The Lorentz Length Contraction

In 1887, A. A. Michelson and E. W. Morley carried out an experiment[1] attempting to detect the "world ether," which was thought to be the invisible medium occupying the entire universe and transmitting electromagnetic effects and radiation. In spite of the great sensitivity of their apparatus, no ether was detected. In an attempt to explain the negative result of the experiment without abandoning the idea of the ether, G. F. Fitzgerald in 1889 and H. A. Lorentz in 1892 proposed a hypothesis[2,3] that, because of an interaction with the ether, all bodies are contracted in the direction of their motion relative to the ether by a factor $(1 - v^2/c^2)^{1/2}$. This hypothesis provides a plausible explanation of the transformation equation for the x coordinate [our Eq. (7-1.1)] in the Lorentz-Poincaré relativity theory, and the effect (albeit hypothetical) became known as "Lorentz contraction."

A. Einstein in his famous 1905 article[4] discarded the idea of world ether as "superfluous" and presented a derivation of the Lorentz transformation equations of coordinates and time on the basis of his postulates of relativity and of independence of the velocity of light on the velocity of the emitter.[5] However, while rejecting the reality of ether, he accepted length contraction of moving bodies as an observable effect, and stated that all moving objects "viewed" from a stationary system appear shortened in the ratio 1 to $(1 - v^2/c^2)^{1/2}$. He also suggested the following method for measuring the length of a moving object (rod): observers in the stationary system ascertain at what points of the stationary system the two ends of the moving rod are located at the same time t; the distance between these two points is the "length of the moving rod." In a later paper Einstein emphasized that this was a measuring procedure fundamentally different from the procedure used for measuring the length of stationary objects.[6] Therefore Einstein's measuring procedure actually constituted the definition of the new quantity, which he called "length of a moving body," different from "length" in the conventional sense.[7] Clearly then, to say that the "length of a moving body" is shorter than the "length" of a body is not the same as to say that the body becomes shorter when it moves. Moreover, it is far from clear how the visual appearance of a moving body can be associated with Einstein's measuring procedure, since the visual appearance is an optical effect unrelated to the measuring procedure proposed by Einstein. It is not surprising therefore that the reality of length contraction and its concrete effect on the appearance of moving bodies has been a subject of considerable controversy and re-evaluation.[8] It should be noted that although Einstein's relativistic length contraction has nothing to do with the world ether, it continues to be known as the "Lorentz contraction."

Taking into account that in Chapters 6 and 7 we obtained correct relativistic transformation equations on the basis of the *retarded* length and volume of moving charge distributions, taking into account that Lorentz contraction requires not one but two observers (two points of observation) for its exact manifestation, and taking into account that electromagnetic fields and light propagate with the same speed, we have hardly any choice but to conclude that the relativistically correct visual shape of a moving body is its retarded shape. We then also have a clear answer to why the retarded field theory, without using Lorentz contraction for determining the effective shape of the moving charge, yields relativistically correct fields of the charge (see Chapter 5 and Sections 7-5 and 7-6). The answer is very simple: as a physical phenomenon Lorentz contraction does not exist. And the fact that the several revisions of this concept had no ill effect on relativistic electrodynamics or on any other branch of physics is an excellent indication that the concept does not represent a physical phenomenon in the conventional sense.

▼

Example 9-1.1 In 1888, on the basis of Maxwell's equations, Oliver Heaviside[9] obtained the equation for the electric field of a point charge q moving with constant velocity v

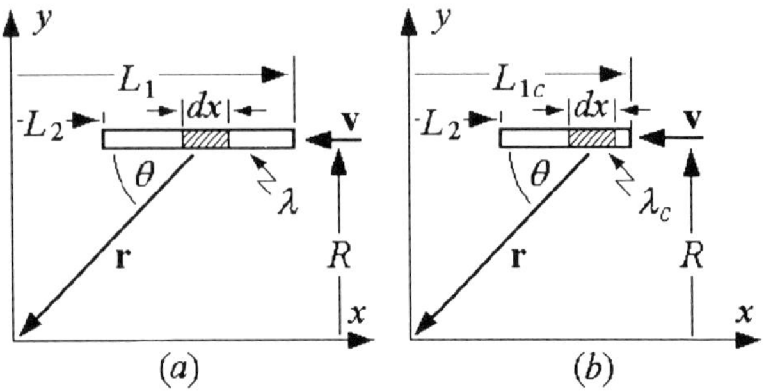

Fig. 9.1 A line charge is moving with velocity $\mathbf{v} = -v\mathbf{i}$. *To obtain the correct expression for the electric field of the charge, one must use the ordinary length of the charge (Fig 9.1a). If the Lorentz-contracted length is used (Fig. 9.1b), the resulting field is incorrect.*

$$\mathbf{E} = \frac{q(1 - v^2/c^2)}{4\pi\varepsilon_0 r^3[1 - (v^2/c^2)\sin^2\theta]^{3/2}}\mathbf{r}, \tag{9-1.1}$$

where $\mathbf{r}$ is the vector connecting the point charge with the point of observation, and θ is the angle between $\mathbf{v}$ and $\mathbf{r}$. We obtained the same equation in Chapter 4 [Eq. (4-1.13)] on the basis of electromagnetic retardation and also in Chapter 7 [Eq. (7-6.21)] on the basis of relativistic transformations. Thus there is no doubt that this equation is correct. In Section 4-3, we found on the basis of electromagnetic retardation the equations for the electric field of a moving line charge of finite length [Eqs. (4-3.13) and (4-3.22)]

$$E_x = \frac{\lambda(1 - v^2/c^2)}{4\pi\varepsilon_0 R}\left[\frac{1}{(L_1^2/R^2 + 1 - v^2/c^2)^{1/2}} - \frac{1}{(L_2^2/R^2 + 1 - v^2/c^2)^{1/2}}\right], \tag{9-1.2}$$

$$E_y = \frac{\lambda}{4\pi\varepsilon_0 R^2}\left[\frac{L_2}{(L_2^2/R^2 + 1 - v^2/c^2)^{1/2}} - \frac{L_1}{(L_1^2/R^2 + 1 - v^2/c^2)^{1/2}}\right]. \tag{9-1.3}$$

In Chapter 7 we obtained the same equations on the basis of

relativistic transformations [Eqs. (7-6.25) and (7-6.27)]. Thus there is no doubt that also these two equations are correct.[10] Show that the same two equations can be obtained by integrating Heaviside's Eq. (9-1.1) over the actual length of the moving charge, but not by integrating over the Lorentz-contracted length (thus demonstrating that Lorentz contraction is not a true physical effect).

Replacing in Eq. (9-1.1) q by λdx and integrating the x component of Eq. (9-1.1) between L_2 and L_1 (see Fig. 9.1), we obtain for E_x (observe that $\mathbf{r}$ is directed toward the point of observation so that its x and y components are negative)

$$\begin{aligned} E_x &= -\frac{\lambda}{4\pi\varepsilon_0}\int_{L_1}^{L_2}\frac{(1-v^2/c^2)}{r^3[1-(v^2/c^2)\sin^2\theta]^{3/2}}xdx \\ &= -\frac{\lambda(1-v^2/c^2)}{4\pi\varepsilon_0}\int_{L_1}^{L_2}\frac{xdx}{[x^2+R^2(1-v^2/c^2)]^{3/2}} \\ &= \frac{\lambda(1-v^2/c^2)}{4\pi\varepsilon_0 R}\left[\frac{1}{(L_1^2/R^2+1-v^2/c^2)^{1/2}}-\frac{1}{(L_2^2/R^2+1-v^2/c^2)^{1/2}}\right]. \end{aligned} \tag{9-1.4}$$

For E_y we similarly obtain

$$\begin{aligned} E_y &= -\frac{\lambda}{4\pi\varepsilon_0}\int_{L_1}^{L_2}\frac{(1-v^2/c^2)}{r^3[1-(v^2/c^2)\sin^2\theta]^{3/2}}Rdx \\ &= -\frac{\lambda(1-v^2/c^2)R}{4\pi\varepsilon_0}\int_{L_1}^{L_2}\frac{dx}{[x^2+R^2(1-v^2/c^2)]^{3/2}} \\ &= \frac{\lambda}{4\pi\varepsilon_0 R^2}\left[\frac{L_2}{(L_2^2/R^2+1-v^2/c^2)^{1/2}}-\frac{L_1}{(L_1^2/R^2+1-v^2/c^2)^{1/2}}\right]. \end{aligned} \tag{9-1.5}$$

These are exactly the same equations as Eqs. (9-1.2) and (9-1.3).

Let us now assume that the line charge is Lorentz contracted. Then its charge density will be not λ but $\lambda_c = \gamma\lambda$ (because the total charge must remain unchanged). Furthermore, if the position of the leading end of the charge is L_2, then the position of the trailing end will be $L_{1c} = L_2 + (L_1 - L_2)/\gamma$. Therefore the Lorentz-contracted versions of Eqs. (9-1.4) and (9-1.5) are

$$E_x = \frac{\lambda_c(1 - v^2/c^2)}{4\pi\varepsilon_0 R}\left[\frac{1}{(L_{1c}^2/R^2 + 1 - v^2/c^2)^{1/2}} - \frac{1}{(L_2^2/R^2 + 1 - v^2/c^2)^{1/2}}\right] \tag{9-1.6}$$

and

$$E_y = \frac{\lambda_c}{4\pi\varepsilon_0 R^2}\left[\frac{L_2}{(L_2^2/R^2 + 1 - v^2/c^2)^{1/2}} - \frac{L_{1c}}{(L_{1c}^2/R^2 + 1 - v^2/c^2)^{1/2}}\right], \tag{9-1.7}$$

which are quite different from the correct Eqs. (9-1.2) and (9-1.3).

▲

9-2. The Electric Field of a Moving Parallel-Plate Capacitor

A typical elementary problem involving electromagnetic transformation equations is the problem of finding the electric field of a parallel-plate capacitor moving with uniform velocity in a direction parallel to its plates. In some textbooks on electromagnetic theory and relativity this problem is solved as follows: "The electric field in the stationary capacitor is $E = \sigma/\varepsilon_0$. If the capacitor is moving, the length of the plates is Lorentz-contracted by the factor γ, so that the surface charge density σ of the capacitor is increased by the factor γ. Therefore the electric field $\mathbf{E}_m$ in the moving capacitor is

$$\mathbf{E}_m = \gamma\mathbf{E}_s, \tag{9-2.1}$$

where $\mathbf{E}_s$ is the electric field in the stationary capacitor."[11]

In some textbooks the same problem is solved as follows: "Let the plates of the capacitor be parallel to the xz plane. The electric field in the stationary capacitor is then $\mathbf{E}_s$. Using the Lorentz-Einstein transformation equation for the y component of the electric field [our Eq. (7-1.6)] and taking into account that there is no magnetic field in the stationary capacitor, we obtain

$$\mathbf{E}_m = \gamma \mathbf{E}_s \tag{9-2.2}$$

for the electric field in the moving capacitor."

Similar solutions are frequently presented for the electric field of an infinitely long line charge uniformly moving in the direction of its length. Invoking the Lorentz length contraction and the corresponding increase of the charge density, or using the transformation equation for the perpendicular component of the electric field, the equation for the electric field of a moving line charge is "shown to be"

$$\mathbf{E}_m = \frac{\gamma\lambda}{2\pi\varepsilon_0 r^2}\mathbf{r} = \gamma \mathbf{E}_s, \tag{9-2.3}$$

where λ is the line charge density, and $\mathbf{r}$ is the vector directed from the line charge to the point of observation at right angles to the line charge.

Clearly, Eq. (9-2.3) is compatible with Eqs. (9-2.1) and (9-2.2) because the electric field of a parallel-plate capacitor can be regarded as a superposition of the electric fields of infinitesimally narrow charged ribbons whose fields ($d\mathbf{E}_m$) are given by Eq. (9-2.3) with $\lambda = \sigma\, dw$, where w is the width of the capacitor plates.

Equation (9-2.3) for the electric field of the moving infinitely long line charge is, however, not at all as obvious as it is claimed to be. In fact, if the electric field of the moving line charge is determined by integrating Heaviside's equation for the electric field of a moving point charge (see Example 9-1.1) the result is[12]

$$\mathbf{E}_m = \frac{\lambda}{2\pi\varepsilon_0 r^2}\mathbf{r} = \mathbf{E}_s, \tag{9-2.4}$$

that is, the field of the moving infinitely long line charge is exactly the same as the field of the same stationary line charge.

But if the correct electric field of a line charge moving along its length is the same as that of the stationary line charge, then also the correct electric field of a parallel-plate capacitor moving in a direction parallel to its plates is the same as the field of the stationary capacitor, that is

$$\mathbf{E}_m = \mathbf{E}_s \tag{9-2.5}$$

rather than the field given by Eq. (9-2.1) or Eq. (9-2.2).

Consider now the magnetic field. According to Eq. (3-2.6), the magnetic flux density field $\mathbf{B}_m$ of any charge distribution moving with uniform velocity $\mathbf{v}$ is connected with the electric field of this distribution by

$$\mathbf{B}_m = (\mathbf{v} \times \mathbf{E}_m)/c^2. \tag{9-2.6}$$

Therefore, if the electric field of the moving capacitor is correctly given by Eqs. (9-2.1) or (9-2.2), then the magnetic flux density field of this capacitor should be [using the vector notation $\mathbf{E}_s = (\sigma/\varepsilon_0)\mathbf{j}$]

$$\mathbf{B}_m = \frac{\gamma\sigma}{c^2\varepsilon_0}\mathbf{v} \times \mathbf{j} = \gamma\mu_0\sigma\mathbf{k}, \tag{9-2.7}$$

where we have used $c^2 = 1/\varepsilon_0\,\mu_0$. Likewise, if the electric field of a moving line charge is correctly given by Eq. (9-2.3), then the magnetic flux density field of this charge should be

$$\mathbf{B}_m = \frac{\gamma\lambda}{c^2 2\pi\varepsilon_0 r^2}\mathbf{v} \times \mathbf{r} = \frac{\gamma\mu_0\lambda}{2\pi r^2}\mathbf{v} \times \mathbf{r}. \tag{9-2.8}$$

But if the electric fields of the moving capacitor and of the line charge are correctly given by Eqs. (9-2.5) and (9-2.4), respectively, then the corresponding magnetic flux density fields should be

$$\mathbf{B}_m = \frac{\sigma}{c^2\varepsilon_0}\mathbf{v} \times \mathbf{j} = \mu_0\sigma v\mathbf{k}, \tag{9-2.9}$$

and

$$\mathbf{B}_m = \frac{\lambda}{c^2 2\pi\varepsilon_0 r^2}\mathbf{v} \times \mathbf{r} = \frac{\mu_0\lambda}{2\pi r^2}\mathbf{v} \times \mathbf{r}. \tag{9-2.10}$$

In Maxwellian electrodynamics the convection current density is defined as ρv, so that the current produced by a line charge moving with velocity v in the direction of its length is λv. By Ampere's law[13], the magnetic field of the moving line charge must then be the field given by Eq. (9-2.10) rather than by Eq. (9-2.8), and therefore the correct electric field of the moving line charge must be the field given by Eq. (9-2.4) rather than by Eq. (9-2.3).

One could argue, of course, as some authors do, that the convection current should be properly defined as $\gamma\rho v$ rather than as ρv. But if one so redefines the convection current, then one must accept that Maxwell's equations themselves are incorrect (because Maxwell's equation for $\nabla\times\mathbf{H}$ involves ρv rather than $\gamma\rho v$). And if one accepts that Maxwell's equations are incorrect, then one must also accept that relativistic electrodynamics is incorrect, since it is based on Maxwell's equations. Thus, unless we are willing to reject the most fundamental relations of both classical and the relativistic electrodynamics, we must conclude that Eq. (9-2.3), and therefore Eqs. (9-2.1) and (9-2.2), are wrong.

Obviously then, the reasoning upon which Eqs. (9-2.1), (9-2.2) and (9-2.3) are based is wrong. In order to understand the fallacy of the arguments leading to Eqs. (9-2.1), (9-2.2), and (9-2.3) it is necessary to look into the origin of the relativistic transformation equations.

The Lorentz transformation equations (see Section 7-1) were first discovered by Lorentz and by other scientists as relations that, when used in the aggregate, made it possible to adopt Maxwell's equations to uniformly moving reference frames without changing the mathematical form of these equations.[14] The same transformation equations were also obtained by Einstein as relations that made Maxwell's equations valid for uniformly moving coordinate systems.[15]

We obtained these equations in Chapters 6 and 7 by considering retarded electric and magnetic fields. Our derivations show very clearly that none of the Lorentz transformation equations can be ignored when transforming electric or magnetic fields from one reference frame to another. Therefore one cannot obtain correct expressions for electric and magnetic fields by means of Lorentz transformations unless all transformable quantities involved in the system under considerations are transformed.

Thus the true significance of the Lorentz transformation equations is not in what this or that individual equation may mean, but in the fact that when taken together, and only when taken together,[16] they constitute an "operator," a "machine," that allows one to convert Maxwell's equations, and therefore all solutions of Maxwell's equations, from one reference frame to another.

Obviously then, none of the relativistic transformation equations may be regarded as ordinary physical equations expressing physical laws or relationships between physical quantities. Relativistic transformation equations must be regarded as prescriptions for replacing one *set* of quantities by another *set* in order to obtain relations between quantities pertaining to one inertial frame of reference from the corresponding relations between quantities pertaining to another inertial frame of reference.

The error in the reasoning leading to Eqs. (9-2.1), (9-2.2), and (9-2.3) is now clear: the equations were obtained by assuming that a *single* relativistic transformation equation had an independent physical validity and by transforming *just one* of the transformable quantities involved in the system under consideration [of course, invoking the non-existent Lorentz contraction for obtaining Eq. (9-2.1) or Eq. (9-2.3) was also wrong, as explained in Section 9-1].

The correct application of relativistic transformations for obtaining the electric and magnetic fields of a moving parallel-plate capacitor is shown in the next section.

9-3. Using Lorentz Transformations for Finding Electric and Magnetic Fields of a Moving Parallel-Plate Capacitor

We shall now show the correct use of Lorentz transformation equations for determining the electric and magnetic fields of the moving capacitor discussed in Section 9-2 (the correct electric and magnetic fields of the moving line charge were already obtained in Section 4-3 as well as in Example 7-6.3 and there is no need to repeat the calculations here).

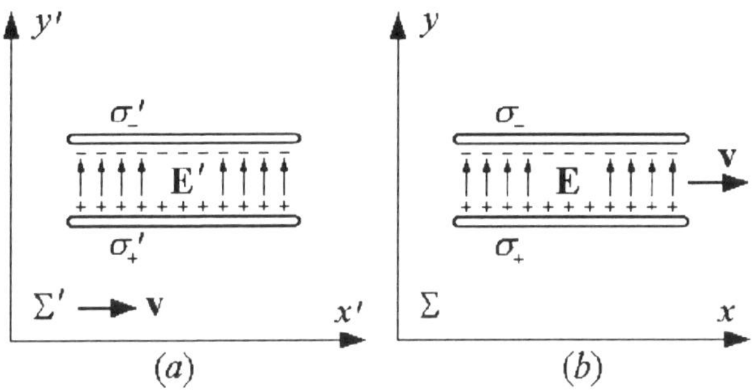

Fig. 9.2 The electric field in a moving parallel-plate capacitor (Fig. 9.2b) is the same as when the capacitor is at rest (Fig. 9.2a). (The distance between the plates is assumed to be very small.)

In the reference frame Σ' co-moving with the capacitor the electric field of the capacitor is[17] (we assume that the capacitor is thin and that its plates are parallel to the *xz* plane, see Fig. 9.2)

$$E_y' = \frac{\sigma'}{\varepsilon_0}. \tag{9-3.1}$$

There are two transformable quantities in this equation: E_y' and σ'. Using Eqs. (7-1.5)-(7-1.7) and (7-1.11), taking into account that $\sigma' = \rho' w$, where w is the thickness of the capacitor's plates [which, by Eq. (7-1.2) is not affected by transformations], and taking into account that there is no current and no magnetic field in the Σ' frame, we obtain for the electric field of the moving capacitor as measured in the laboratory

$$\frac{E_y}{\gamma} = \frac{\sigma}{\gamma \varepsilon_0} \tag{9-3.2}$$

or

$$E_y = \frac{\sigma}{\varepsilon_0}, \tag{9-3.3}$$

which is exactly the same field as in the stationary capacitor. In the vector notation we then have

$$\mathbf{E}_m = \mathbf{E}_s, \tag{9-3.4}$$

in agreement with Eq. (9-2.5).

Using Eqs. (7-1.8)-(7-1.10) and taking into account that $E_z' = 0$ and that there is no magnetic field in the Σ' frame, and then using Eq. (7-1.6), we obtain for the magnetic flux density field of the moving capacitor as measured in the laboratory

$$\mathbf{B} = \frac{\gamma v E_y'}{c^2}\mathbf{k} = \frac{\gamma v E_y}{\gamma c^2}\mathbf{k} = \mu_0 \sigma v \mathbf{k}, \tag{9-3.5}$$

which is the same as Eq. (9-2.9) (we omit the subscript "m" here as superfluous).

9-4. The Right-Angle Lever Paradox

Numerous "relativistic paradoxes" can be found in the literature on relativity. They usually reflect a lack of understanding of the physical significance of relativistic equations. One of the oldest of such paradoxes is the so-called "right-angle lever," or "L-shaped lever paradox," also known as the "Lewis-Tolman lever paradox." It was first reported in 1909.[18]

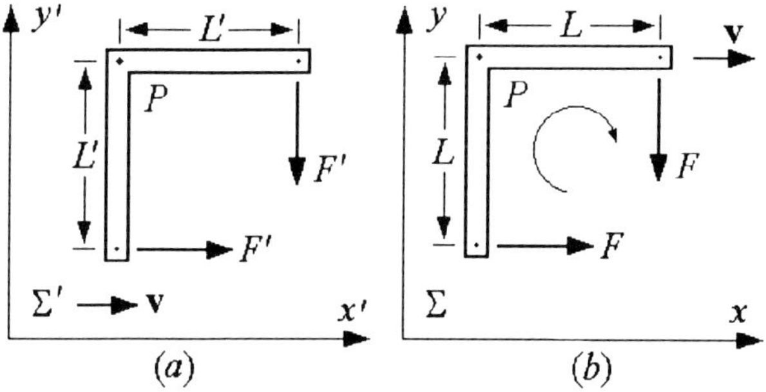

Fig. 9.3 The right-angle lever is in equilibrium in the moving reference frame (Fig. 9.3a) but, according to seemingly correct calculations, should be rotating when observed in the stationary reference frame (Fig. 9.3b).

The essence of the paradox is as follows. Consider an L-shaped lever at rest in the reference frame Σ' moving with velocity v relative to the laboratory frame Σ (Fig. 9.3). Two equal forces F' are applied to the lever at right angles to the arms and at equal distances L' from the pivot P. Since

$$L_x'F_y' = L_y'F_x' = L'F', \qquad (9\text{-}4.1)$$

the torque is

$$T' = L_x'F_y' - L_y'F_x' = 0, \qquad (9\text{-}4.2)$$

so that the lever is in equilibrium in Σ'. Using now Eqs. (8-6.11)-(8-6.13) to transform the torque to the laboratory frame Σ and substituting $T_z' = 0$, we obtain

$$T_z = -(v^2/c^2)r_x'F_y = -(v^2/c^2)L_x'F_y'. \qquad (9\text{-}4.3)$$

Thus in the laboratory frame Σ the lever experiences a net torque and appears to be not in equilibrium. This result is considered to be a paradox, because by the principle of relativity, if a physical system is in equilibrium in one inertial reference frame, it is in equilibrium when observed in any other inertial reference frame.

Numerous articles in scientific journals have been devoted to this paradox proposing a variety of solutions of ever increasing complexity,[19] and many books on relativity have described the paradox without arriving at a meaningful solution. Some authors, unable to present an acceptable solution, prefer to leave the paradox unsolved. Thus, for example, after explaining the paradox, the author of an authoritative and comprehensive book on special relativity theory concludes by saying: "We shall let the reader contemplate about it."

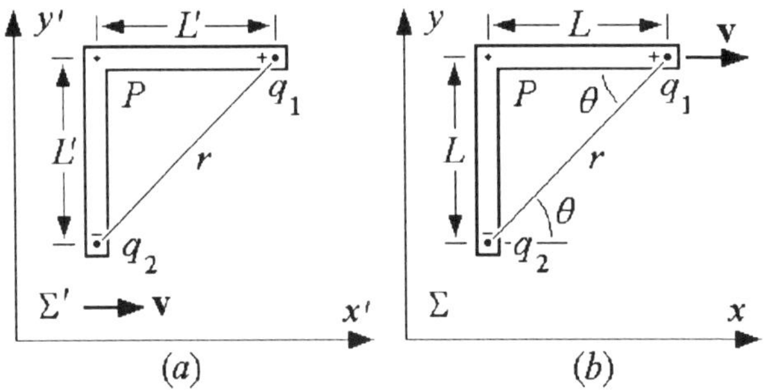

Fig. 9.4 The right-angle lever paradox can be resolved if instead of the unspecified abstract forces one uses real physical forces, such as the forces created by two interacting opposite electric charges.

To reveal the error in the reasoning leading to this paradox, let us consider the system shown in Fig. 9.4. This system is similar to the one shown in Fig. 9.3, except that instead of the two undefined forces applied to the lever, the forces applied to the lever are now caused by two equal and opposite electric charges q_1 and q_2 placed on the two arms of the lever at equal distances L from the pivot. In the Σ' reference frame, the forces between the charges are purely electrostatic, each charge exerting on the other a force of the same magnitude but in opposite direction

$$\mathbf{F}' = \frac{q_1 q_2}{4\pi\varepsilon_0 r'^3}\mathbf{r}', \tag{9-4.4}$$

where r' is the distance between the charges, and $\mathbf{r}'$ is directed from the field-producing to the field-experiencing charge. By the symmetry of the system, the net force and torque acting on the lever is zero.

Consider now the same lever with the two charges as observed in the laboratory reference frame Σ. Let us first analyze the lever in terms of classical electrodynamics without any reference to the relativity theory. Since the lever with the charges moves with respect to the laboratory, the electric field produced by each of the two charges is now according to Heaviside's Eq. (4-1.13)

$$\mathbf{E} = \frac{q(1 - v^2/c^2)}{4\pi\varepsilon_0 r^3[1 - (v^2/c^2)\sin^2\theta]^{3/2}}\mathbf{r}, \tag{9-4.5}$$

where θ is the angle between the velocity vector of the field-producing charge $\mathbf{v}$ and vector $\mathbf{r}$ directed from the field-producing to the field-experiencing charge. The electric force exerted by the charges one upon the other is now

$$\mathbf{F} = \frac{q_1 q_2(1 - v^2/c^2)}{4\pi\varepsilon_0 r^3[1 - (v^2/c^2)\sin^2\theta]^{3/2}}\mathbf{r}. \tag{9-4.6}$$

Again, by the symmetry of the system, the net force and the net torque experienced by the lever due to electric interaction between the charges is zero.

However, a moving charge produces a magnetic field, and if a charge moves in a magnetic field it experiences a Lorentz force. Thus, in the laboratory reference frame not only the electric interaction between the charges but also the magnetic interaction must be taken into account.

Consider the magnetic field produced by the charge q_1 located in the horizontal arm of the lever. The charge moves with velocity $\mathbf{v} = v\mathbf{i}$. According to Eqs. (3-2.6) and (9-4.5), the magnetic field (as measured in the laboratory) produced by this charge at the point where the charge q_2 is located is

$$\mathbf{B}_1 = \frac{q_1(1 - v^2/c^2)}{4\pi\varepsilon_0 c^2 r^3[1 - (v^2/c^2)\sin^2\theta]^{3/2}} \mathbf{v} \times \mathbf{r}, \tag{9-4.7}$$

where $\mathbf{r}$ is directed from q_1 to q_2.

Since the charge q_2 moves through this field with velocity $\mathbf{v} = v\mathbf{i}$, it experiences a magnetic force $\mathbf{F}_2 = q_2\mathbf{v} \times \mathbf{B}_1$, or

$$\mathbf{F}_2 = \frac{q_1 q_2(1 - v^2/c^2)}{4\pi\varepsilon_0 c^2 r^3[1 - (v^2/c^2)\sin^2\theta]^{3/2}} \mathbf{v} \times (\mathbf{v} \times \mathbf{r}). \tag{9-4.8}$$

Expanding the triple cross product in Eq. (9-4.8) by means of vector identity (V-3), we have

$$\mathbf{v} \times (\mathbf{v} \times \mathbf{r}) = \mathbf{v}(\mathbf{v} \cdot \mathbf{r}) - \mathbf{r}(\mathbf{v} \cdot \mathbf{v}) = \mathbf{v}(\mathbf{v} \cdot \mathbf{r}) - v^2\mathbf{r}, \tag{9-4.9}$$

and if we express $\mathbf{r}$ in terms of its Cartesian components as $\mathbf{r} = -L\mathbf{i} - L\mathbf{j}$ and use $\mathbf{v} = v\mathbf{i}$, we find that the triple product reduces to

$$\mathbf{v} \times (\mathbf{v} \times \mathbf{r}) = v\mathbf{i}[v\mathbf{i} \cdot (-L\mathbf{i} - L\mathbf{j})] - v^2(-L\mathbf{i} - L\mathbf{j}) = v^2 L\mathbf{j}. \tag{9-4.10}$$

The force experienced by the charge q_2 due to the magnetic field produced by the charge q_1 is therefore

$$\mathbf{F}_2 = \left(\frac{v^2}{c^2}\right) \frac{q_1 q_2(1 - v^2/c^2)L}{4\pi\varepsilon_0 r^3[1 - (v^2/c^2)\sin^2\theta]^{3/2}} \mathbf{j}. \tag{9-4.11}$$

Since this force is directed along the vertical lever arm, it produces no torque on the lever.

Consider now the magnetic field produced by the charge q_2 located in the vertical arm of the lever. The charge q_2 moves with velocity $\mathbf{v} = v\mathbf{i}$. According to Eqs. (3-2.6) and (9-4.5), the magnetic field produced by this charge at the point where the charge q_1 is located is

$$\mathbf{B}_2 = \frac{q_1(1 - v^2/c^2)}{4\pi\varepsilon_0 c^2 r^3[1 - (v^2/c^2)\sin^2\theta]^{3/2}} \mathbf{v} \times \mathbf{r}, \qquad (9\text{-}4.12)$$

where $\mathbf{r}$ is directed from q_2 to q_1.

The charge q_1 moves through this field with velocity $\mathbf{v} = v\mathbf{i}$ and therefore experiences a magnetic force $\mathbf{F}_1 = q_1\mathbf{v} \times \mathbf{B}_2$, or

$$\mathbf{F}_1 = \frac{q_1 q_2(1 - v^2/c^2)}{4\pi\varepsilon_0 c^2 r^3[1 - (v^2/c^2)\sin^2\theta]^{3/2}} \mathbf{v} \times (\mathbf{v} \times \mathbf{r}). \qquad (9\text{-}4.13)$$

The triple cross product in Eq. (9-4.13) is the same as in Eq. (9-4.9), and if we express $\mathbf{r}$ in terms of its Cartesian components as $\mathbf{r} = L\mathbf{i} + L\mathbf{j}$ and use $\mathbf{v} = v\mathbf{i}$, we find that the triple product reduces to

$$\mathbf{v} \times (\mathbf{v} \times \mathbf{r}) = v\mathbf{i}[v\mathbf{i} \cdot (L\mathbf{i} + L\mathbf{j})] - v^2(L\mathbf{i} + L\mathbf{j}) = -v^2 L\mathbf{j}. \qquad (9\text{-}4.14)$$

The force experienced by the charge q_1 due to the magnetic field produced by the charge q_2 is therefore

$$\mathbf{F}_1 = -\left(\frac{v^2}{c^2}\right) \frac{q_1 q_2(1 - v^2/c^2)L}{4\pi\varepsilon_0 r^3[1 - (v^2/c^2)\sin^2\theta]^{3/2}} \mathbf{j}. \qquad (9\text{-}4.15)$$

This force is perpendicular to the horizontal lever arm and therefore produces a torque

$$\mathbf{T} = \mathbf{r} \times \mathbf{F} = -\left(\frac{v^2}{c^2}\right) \frac{q_1 q_2(1 - v^2/c^2)L^2}{4\pi\varepsilon_0 r^3[1 - (v^2/c^2)\sin^2\theta]^{3/2}} \mathbf{k}. \qquad (9\text{-}4.16)$$

Thus, the appearance of the torque on the moving lever carrying the two charges is an electromagnetic rather than a relativistic effect.

In terms of the electromagnetic theory, there are just two possibilities to explain the appearance of the torque: either there is a "hidden" torque equalizing the torque given by Eq. (9-4.16), or at the same time as the torque acts on the lever, the lever increases its angular momentum at the rate and in the direction that counteracts the effect of the torque. Since we cannot find a hidden torque, let us explore the latter possibility.

According to Eq. (8-2.3), whenever a point charge q is located in the magnetic field whose magnetic vector potential is $\mathbf{A}$, there exists an electromagnetic momentum

$$\mathbf{G} = q\mathbf{A}, \tag{9-4.17}$$

associated with this charge. Therefore the two charges in the lever are associated with electromagnetic momenta, and each charge is associated also with an angular electromagnetic momentum given by

$$\mathbf{I} = \mathbf{L}\times\mathbf{G} = q\mathbf{L}\times\mathbf{A}, \tag{9-4.18}$$

where $\mathbf{L}$ is the vector (lever arm) joining the pivot with the charge under consideration.

The vector potential at the location of q_1 is produced by the charge q_2 and, according to Eq. (4-5.6), is parallel to $\mathbf{v}$. Since the lever arm of the charge q_1 is parallel to the x axis, and since $\mathbf{v}$ is also parallel to the x axis, the angular electromagnetic momentum associated with q_1 is, by Eq. (9-4.19), zero.

The vector potential at the location of q_2 is produced by the charge q_1 and is in the direction of the velocity vector $\mathbf{v}$. Since the lever arm of the charge q_2 is along the y axis and since its magnitude is L, the angular momentum associated with q_2 is, by Eq. (9-4.18),

$$\mathbf{I}_2 = q_2 L A_1 \mathbf{k}, \tag{9-4.19}$$

where A_1 is the magnetic vector potential produced by the charge q_1 at the location of the charge q_2.

As the lever moves, its position changes, and so does the location of the magnetic angular momentum. The time rate of change of the electromagnetic angular momentum is given by the substantive derivative

$$\frac{\partial \mathbf{I}_2}{\partial t} = -\mathbf{k} q_2 L(\mathbf{v} \cdot \nabla) A_1 = -\mathbf{k} q_2 L v \frac{\partial A_1}{\partial x}. \qquad (9\text{-}4.20)$$

Now, by Eqs. (4-5.5) and (4-5.6),

$$A_1 = \varepsilon_0 \mu_0 v \varphi_1 = (v/c^2)\varphi_1, \qquad (9\text{-}4.21)$$

where φ_1 is the electric scalar potential produced by q_1 at the location of q_2. Equation (9-4.20) becomes therefore

$$\frac{\partial \mathbf{I}_2}{\partial t} = -\mathbf{k} q_1 L\left(\frac{v^2}{c^2}\right)\frac{\partial \varphi_1}{\partial x}. \qquad (9\text{-}4.22)$$

But $-\partial\varphi_1/\partial x$ is the x component of the electric field produced by q_1. By Eq. (9-4.5) and Fig. 9.4 this component is

$$E_{x1} = -\frac{q_1(1 - v^2/c^2)L}{4\pi\varepsilon_0 r^3[1 - (v^2/c^2)\sin^2\theta]^{3/2}}. \qquad (9\text{-}4.23)$$

Combining Eqs. (9-4.22) and (9-4.23) we obtain for the rate of change of the magnetic angular momentum associated with the moving lever

$$\frac{\partial \mathbf{I}_2}{\partial t} = \left(\frac{v^2}{c^2}\right)\frac{q_1 q_2(1 - v^2/c^2)L^2}{4\pi\varepsilon_0 r^3[1 - (v^2/c^2)\sin^2\theta]^{3/2}}\mathbf{k}, \qquad (9\text{-}4.24)$$

which is equal in magnitude and opposite in direction to the torque given by Eq. (9-4.16).

In terms of classical electrodynamics the paradox is now resolved. It is clear that without the torque the angular momentum of the system would increase indefinitely in violation of the law of conservation of angular momentum. With the torque present, the sum of the mechanical angular momentum (due to the torque) and the magnetic angular momentum of the system remains zero. The lever remains in equilibrium because the entire torque is needed to balance the increase in the magnetic angular momentum of the system, and nothing is left for turning the lever.

Let us now analyze the lever with the two charges by means of relativistic transformation.

Assuming for simplicity that the vertical arm of the lever is on the y axis and that the charge q_2 is at the origin of coordinates, we have, by Eq. (9-4.4), for the vertical component of the force experienced by the charge q_1 in the moving reference frame Σ'

$$F'_{y1} = \frac{q_1 q_2}{4\pi\varepsilon_0 r'^3} y' = \frac{q_1 q_2}{4\pi\varepsilon_0 (x'^2 + y'^2)^{3/2}} y', \qquad (9\text{-}4.25)$$

where x' and y' are the coordinates of q_1. By Eq. (9-4.3) the torque experienced by the lever in Σ can then be written as,

$$T_z = -\left(\frac{v^2}{c^2}\right) \frac{q_1 q_2 x' y'}{4\pi\varepsilon_0 (x'^2 + y'^2)^{3/2}}. \qquad (9\text{-}4.26)$$

Transforming x' and y' to the corresponding values in Σ by means of Eqs. (7-1.19) and (7-1.20) with $t = 0$ (we are free to choose the time of observation in Σ because the lever is stationary in Σ'), we obtain

$$T_z = -\left(\frac{v^2}{c^2}\right) \frac{q_1 q_2 \gamma x y}{4\pi\varepsilon_0 (\gamma^2 x^2 + y^2)^{3/2}} = -\left(\frac{v^2}{c^2}\right) \frac{q_1 q_2 \gamma x y}{4\pi\varepsilon_0 \gamma^3 (x^2 + y^2/\gamma^2)^{3/2}}. \qquad (9\text{-}4.27)$$

Expanding γ, using $x = L$, $y = L$, $x^2 + y^2 = r^2$, $y^2/(x^2 + y^2) = \sin^2\theta$ and using vector notation, we can write Eq. (9-4.27) as

$$\mathbf{T} = -\left(\frac{v^2}{c^2}\right) \frac{q_1 q_2 (1 - v^2/c^2) L^2}{4\pi\varepsilon_0 r^3 [1 - (v^2/c^2)\sin^2\theta]^{3/2}} \mathbf{k}, \qquad (9\text{-}4.28)$$

which is exactly the same torque as that given by Eq. (9-4.16) obtained by direct electromagnetic calculations.

Proceeding now as in the case of direct electromagnetic calculations we find as before that the magnetic angular momentum associated with q_1 is zero.

For the magnetic angular momentum associated with q_2 we have

$$\mathbf{I}_2 = q_2 L A_1 \mathbf{k}. \tag{9-4.29}$$

The time rate of change of the electromagnetic angular momentum is

$$\frac{\partial \mathbf{I}_2}{\partial t} = -\mathbf{k} q_2 L v \frac{\partial A_1}{\partial x} = -\mathbf{k} q_2 L \left(\frac{v^2}{c^2}\right) \frac{\partial \varphi_1}{\partial x}. \tag{9-4.30}$$

Since $\partial \varphi_1 / \partial x$ is E_{x1}, we obtain

$$\frac{\partial \mathbf{I}_2}{\partial t} = -\mathbf{k} q_2 L \left(\frac{v^2}{c^2}\right) E_{x1}. \tag{9-4.31}$$

But E_{x1} at the location of q_2 is produced by q_1 and according to Heaviside's Eq. (9-1.1) [or Eq. (7-6.21) or Eq. (7-5.20)] is

$$E_{x2} = -\frac{q_1(1 - v^2/c^2)L}{4\pi\varepsilon_0 r^3[1 - (v^2/c^2)\sin^2\theta]^{3/2}}. \tag{9-4.32}$$

Substituting Eq. (9-4.32) into Eq. (9-4.31) we obtain

$$\frac{\partial \mathbf{I}_2}{\partial t} = \left(\frac{v^2}{c^2}\right) \frac{q_1 q_2 (1 - v^2/c^2) L^2}{4\pi\varepsilon_0 r^3[1 - (v^2/c^2)\sin^2\theta]^{3/2}} \mathbf{k}, \tag{9-4.33}$$

which is equal in magnitude and opposite in direction to the torque given by Eq. (9-4.28). Thus once again we find that the lever is in equilibrium in the laboratory just as it is in the moving reference frame Σ'.

Clearly, the "paradox" is merely a result of an incomplete statement of the problem, when instead of real physical forces one uses some abstract forces F_1 and F_2 (Fig. 9.3) and thus fails to take into account the physical effects that take place when the forces are created by well-defined physical interactions.

9-5. Is the Magnetic Field due to an Electric Current a Relativistic Effect?

Several authors have asserted that the magnetic field due to an electric current is a relativistic effect. This assertion is based on the fact that if one assumes that the interaction between electric charges is entirely due to the electric field, then relativistic field or force transformation equations demand the existence of the magnetic field.[20]

It is shown below that one could assert with equal justification that the electric field rather than the magnetic field is a relativistic effect. Therefore, since it is impossible for both fields to be relativistic effects, neither field should be regarded as a relativistic effect.[21]

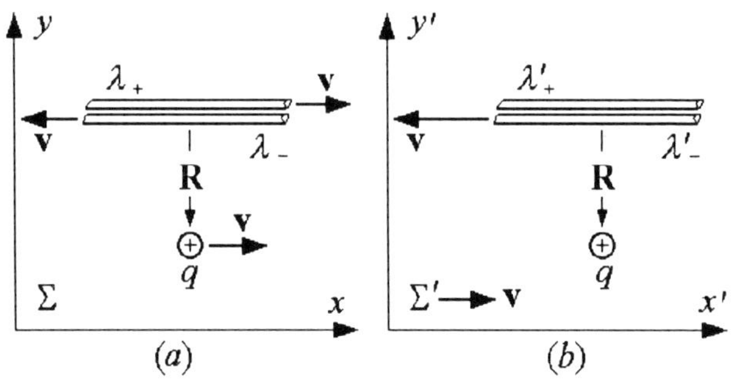

Fig. 9.5 In the laboratory reference frame Σ (Fig. 9.5a) the point charge q experiences a magnetic force. But in the moving reference frame Σ' (Fig. 9.5b) the charge experiences an electric force.

Consider two very long ("infinitely long") line charges of opposite polarity adjacent to each other along their entire length and observed in the laboratory reference frame Σ. Let the charges be parallel to the x axis and let the magnitude of the line charge density in each line charge be λ. Let the positive line charge move with velocity $\mathbf{v} = v\mathbf{i}$ and let the negative line charge move with velocity $\mathbf{v} = -v\mathbf{i}$ (Fig. 9.5). Let us now assume that a positive point charge q is present in the xy plane at a distance R from the line charges and let us assume that it moves with velocity $\mathbf{v}$ in the positive direction of the x axis.

In the laboratory reference frame Σ the two line charges constitute a current $2\lambda v$. By Ampere's law,[13] the magnetic flux density field that this current produces at the location of q is

$$\mathbf{B} = \mu_0 \frac{\lambda \mathbf{v} \times \mathbf{R}}{\pi R^2}, \tag{9-5.1}$$

where $\mathbf{v} = v\mathbf{i}$, and where $\mathbf{R}$ is the vector joining the line charge with q and directed toward q. The force exerted by $\mathbf{B}$ on q is

$$\mathbf{F} = q(\mathbf{v} \times \mathbf{B}) = q\left(\mathbf{v} \times \mu_0 \frac{\lambda \mathbf{v} \times \mathbf{R}}{\pi R^2}\right), \tag{9-5.2}$$

or

$$\mathbf{F} = -\mu_0 \frac{q\lambda v^2}{\pi R^2}\mathbf{R}. \tag{9-5.3}$$

Let us now look at the two line charges and the point charge from a reference frame Σ' moving with velocity $\mathbf{v} = v\mathbf{i}$ relative to the laboratory. The point charge q is stationary in this reference frame and therefore experiences no magnetic force at all.

However, according to the relativistic force transformation equations [Eqs. (8-1.14), (8-1.17) and (8-1.18) with $u_x' = u_y' = u_z' = 0$], if q experiences a radial force $\mathbf{F}$ (in the y direction) in the laboratory reference frame Σ, then it must experience a radial force $\mathbf{F}'$ in the moving reference frame Σ' given by

$$\mathbf{F}' = \mathbf{F}(1 - v^2/c^2)^{-1/2}, \tag{9-5.4}$$

or, with Eq. (9-5.3), by

$$\mathbf{F}' = -\mu_0 \frac{q\lambda v^2}{\pi R^2(1 - v^2/c^2)^{1/2}}\mathbf{R} \tag{9-5.5}$$

[$\mathbf{R}$ is the same in both reference frames because of Eq. (7-1.2)].

Of course, Eq. (9-5.5) is not really meaningful unless λ in it is converted to λ' pertaining to the moving reference frame Σ'. To find λ' we use Eq. (7-1.29)

$$\rho' = \gamma[\rho - (v/c^2)J_x]. \tag{9-5.6}$$

The charge density ρ in the laboratory reference frame is $\rho = (\lambda_+ + \lambda_-)/S = 0$, and the current density is $J_x = 2\lambda v/S$, where S is the cross-sectional area of the positive and the negative line charge. Substituting ρ and J_x into Eq. (9-5.6) and multiplying by S, we have

$$\lambda' = -\gamma\frac{2\lambda v^2}{c^2} = -\frac{2\lambda v^2}{c^2(1-v^2/c^2)^{1/2}}. \tag{9-5.7}$$

Substituting Eq. (9-5.7) into Eq. (9-5.5), we obtain for the force acting on the point charge q in the moving reference frame Σ'

$$\mathbf{F}' = \mu_0\frac{c^2 q\lambda'}{2\pi R^2}\mathbf{R}, \tag{9-5.8}$$

and, since $\mu_0 c^2 = 1/\varepsilon_0$,

$$\mathbf{F}' = \frac{q\lambda'}{2\pi\varepsilon_0 R^2}\mathbf{R}, \tag{9-5.9}$$

which is exactly what we would have obtained for the force exerted on q in Σ' by the electric field due to the line charge of density λ' (note that λ' is *negative*, so that the field is directed toward the two line charges).[22]

As is clear from Eqs. (9-5.1)-(9-5.9), relativistic force transformation equations demand the presence of an electric field when the interactions between electric charges are assumed to be entirely due to a magnetic force. We could interpret this result as evidence that the electric field is a relativistic effect. But the well-known fact[20] that similar calculations demand the presence of a magnetic field, if the interactions between the charges are assumed to be entirely due to an electric force, makes such an interpretation impossible (unless we are willing to classify both the magnetic and the electric field as relativistic effects, which is absurd). We must conclude therefore that neither the magnetic nor the electric field is a relativistic effect.

The only correct interpretation of the above calculations must then be that interactions between electric charges that are either entirely velocity independent or entirely velocity dependent is incompatible with the relativity theory. Both fields — the electric field (producing a force *independent* of the velocity of the charge experiencing the force) and the magnetic field (producing a force *dependent* on the velocity of the charge experiencing the force) — are necessary to make interactions between electric charges relativistically correct. By inference then, any force field compatible with relativity theory must have an electric-like "subfield" and a magnetic-like "subfield." In fact, as is shown in Chapter 11, this is exactly what happens in the case of gravitational fields.

References and Remarks for Chapter 9

1. A. A. Michelson and E. W. Morley, "On the Relative Motion of the Earth and the Luminiferous Ether," Amer. J. Sci. **34**, 333-345 (1887). For a discussion see, for example, A. P. French, *Special Relativity* (Norton, New York, 1968) pp. 37-58.
2. G. F. Fitzgerald, "The Ether and the Earth's Atmosphere," Science **13**, 390 (1889). An interesting account of the history of Fitzgerald's hypothesis is given in A. M. Bork "The Firzgerald Contraction Hypothesis," ISIS **57**, 199-207 (1966) and in S. G. Brush, "Note on the History of the Fitzgerald-Lorentz Contraction," ISIS **58**, 230-232 (1967).
3. H. A. Lorentz, "The relative Motion of the Earth and the Ether," Versl. Kon. Acad. Wetensch. Amsterdam **1**, 74-78 (1892).

4. A. Einstein, "Zur Elektrodynamik bewegter Körper," Ann. Phys. **17**, 891-921 (1905).

5. The significance and validity of Einstein's derivation of coordinate and time transformation equations is questionable. According to the authoritative and highly regarded book by A. I. Miller, *Albert Einstein's Special Theory of Relativity* (Addison-Wesley, Reading, Massachusetts, 1991) p. 216, "In summary, on the basis of this chapter it seems as if Einstein knew beforehand the spatial portion of the relativistic transformation and an approximate version of the correct time coordinate. . . It is difficult to imagine that Einstein first derived the relativistic transformations by the method described in the 1905 paper; in fact, he never used this method again."

6. A. Einstein "Die Relativitätstheorie" in E. Lecher, ed., *Physik*, 2nd ed., (Teubner, Leipzig, 1925) p. 791.

7. It may be noted that Einstein's procedure for measuring the length of a moving body cannot actually be implemented. Since neither the trajectory nor the length of a moving body is known beforehand, the procedure requires that observers with clocks should be placed in each and every point of space, which is clearly impossible; moreover, to measure the lengths of moving microscopic particles (electrons, for example) the observers and the clocks would have to be of subatomic dimensions; etc, etc.

8. See, for example, J. Terrell, "Invisibility of the Lorentz Contraction," Phys. Rev. **116**, 1014-1045 (1959); Roy Weinstein, "Observation of length by a single observer," Am. J. Phys. **28**, 607-610 (1960); V. F. Weisskopf, "The visual appearance of rapidly moving objects," Phys. Today **13**, 24-27 (1960); A. Gamba, "Physical quantities in different reference systems according to relativity," Am. J. Phys. **35**, 83-89 (1967); G. D. Scott and M. R. Viner, "The geometrical appearance of large objects moving at relativistic speeds," Am. J. Phys. **33**, 534-536 (1965); V. N. Strel'tsov, "On the relativistic length," Found. Phys. **6** 293-8 (1976); Kevin G. Suffern, "The apparent shape of a moving sphere," Am. J. Phys. **56**, 729-733 (1988); V. N. Strel'tsov, "The Question is: Are Fast-Moving Scales Contracted or Elongated?," Hardonic J. **17**, 105-114 (1994).

9. Oliver Heaviside, "The Electromagnetic Effects of a Moving Charge," The Electrician **22**, 147-148 (1888); Oliver Heaviside, "On the Electromagnetic Effects due to the Motion of Electricity Through a Dielectric," Phil. Mag. **27**, 324-339 (1889). Observe that Heaviside's equation is relativistically correct [see Eqs. (7-5.20), (7-5.31), (7-6.21)]. The first relativistic derivation of the electric field of a point charge was done by H. Poincaré in "Sur la dynamique de l'électron," Ren. Circ. Mat. Palermo **21**, 129-175 (1906).
10. See also Oleg D. Jefimenko, "Retardation and relativity: The case of a moving line charge," Am. J. Phys. **63**, 454-459 (1995).
11. It may be noted that as a sequel to this solution some authors then use Eq. (9-2.1) for "deriving" the Lorentz-Einstein transformation equation for the perpendicular component of the electric field.
12. See, for example, A. P. French, *Special Relativity* (Norton, New York, 1968) pp. 250-253. The same result is obtained from Eq. (9-1.5) by assuming that L_2 is negative and that $|L_2|$, $L_1 \gg R$. In this case $1-v^2/c^2$ in the denominator of Eq. (9-1.5) is negligible, and Eq. (9-1.5) reduces to Eq. (9-2.4). See also Example 5-3.2.
13. See, for example, Oleg D. Jefimenko, *Electricity and Magnetism*, 2nd ed., (Electret Scientific, Star City, 1989) p. 328-332.
14. See Chapter 7, Refs. 1, 2, 6.
15. See Ref. 4.
16. However, certain surrogates, such as relativistic force transformation equations, can sometimes be used in lieu of the Lorentz field transformation equations.
17. See, for example, Ref. 13, p. 114.
18. G. N. Lewis and R. C. Tolman, "The Principle of Relativity, and Non-Newtonian Mechanics," Philos. Mag. **18**, 510-523 (1909).
19. Among the proposed solutions are: to redefine the torque, to invoke elastic forces in the lever, to reexamine the relation between torque and angular momentum, to invoke energy flow in the lever, to redefine the concept of force, etc. See J. W. Butler, "The Lewis-Tolman Lever Paradox," Am. J. Phys. **38**, 360-368 (1970); J. Charles Nickerson and Robert T. McAdory, "Right angle paradox," Am. J. Phys. **43**, 615-621 (1975); G. Cavalleri, Ø. Grøn, and G.

Spinelli, "Comment on the article 'Right-angle level paradox' by J. C. Nickerson and R. T. McAdory," Am. J. Phys. **46**, 108-109 (1978); D. Garth Jensen, "The paradox of the L-shaped object," Am. J. Phys. **57**, 553-555 (1989).

20. See, for example, E. M. Purcell, *Electricity and Magnetism*, 2nd ed., (McGraw-Hill, New York, 1985) pp. 192-196 or V. D. Barger and M. G. Olsson, *Classical Electricity and Magnetism* (Allyn and Bacon, Boston, 1987) pp. 513-515.

21. See also Oleg D. Jefimenko, "Is magnetic field due to an electric current a relativistic effect?," Eur. J. Phys. **17**, 180-182 (1996).

22. See, for example, Ref. 13, pp. 89-90 and 98-99.

10

THE RATE OF MOVING CLOCKS

One of the most enduring relativistic paradoxes is the so-called "clock paradox" (commonly known as the "twin paradox"), according to which time runs slower in moving reference frames than in stationary reference frames. This "time dilation" is considered to be a purely kinematic relativistic effect, a consequence of nothing more than relative motion. Several experiments appear to support the reality of time dilation. However, in the preceding chapters we saw that certain electromagnetic and mechanical interactions between moving bodies are easily overlooked because of their subtleness and their difference from the familiar interactions between stationary bodies. It is conceivable therefore that moving clocks may run slower than stationary clocks as a result of some heretofore ignored interactions affecting moving clocks, rather than as a result of their motion as such. With this idea in mind, we shall compare in this chapter the rates of some primitive electromagnetic "clocks" resting in the laboratory with the rates of the same "clocks" moving with respect to the laboratory by using well-established laws of electromagnetism and mechanics.

10-1. The Idea of Time Dilation

The idea that some physical phenomena occur at a slower rate when the system in which the phenomena take place is moving with respect to the observer dates back to 1897, when Joseph Larmor, using transformations for length and time analogous to Lorentz transformations, concluded that the periods of orbiting electrons are shorter by the factor γ in the rest system than in the moving system.[1] Albert Einstein in his famous 1905 paper interpreted the Lorentz transformation equation of coordinates and time as indicating that the rate of a moving clock, "when viewed from the stationary system," is slower by the factor γ than the rate of the same clock at rest in the stationary system.[2] Later he generalized this statement by declaring that "a living organism after any lengthy flight could be returned to its original spot in a scarcely altered condition, while corresponding organism which had remained in the original position had already long since given way to new generations" and "every happening in a physical system slows down when this system is set in translational motion."[3,4] Thus, according to Einstein, not only clocks run slow, but time itself is "dilated" in systems that move with respect to the systems considered to be stationary (laboratory).

The idea of the slowing down of moving clocks as a strictly kinematic effect was unacceptable to many of Einstein's contemporaries[5] and the idea of time dilation remains to this day one of the most controversial aspects of Einstein's special relativity theory[6]. However, experiments on the radioactive decay of fast mesons show that their decay occurs indeed at a rate slower by the factor γ (within experimental errors) than for resting or slowly-moving mesons.[7,8]

As a physical entity, time is defined in terms of specific measurement procedures, which may be described simply as "observing the rate of clocks." But a clock is a physical apparatus or device and is subject to the laws of physics in accordance with which the clock is constructed. Therefore, if a clock slows down when it moves, its slower rate should be explainable on the basis of the specific laws responsible for the operation of the clock. For some inexplicable reason, apparently nobody has attempted to calculate and compare the rates of any types of stationary and moving clocks, although such a calculation would be of the utmost significance as a means of resolving the above-mentioned controversy and as an answer to the question of whether or not the slow rate of moving clocks (if it can be confirmed by calculations) can be explained as a dynamic cause-and-effect phenomenon rather than as the kinematic effect enunciated by Einstein.

Naturally, insofar as, according to Einstein, the slowing down is supposed to hold for any clock mechanism whatsoever, an all-inclusive dynamic (causal) interpretation of the slow rate of moving clocks is hardly possible. But it should be possible to provide a causal interpretation of the slow rate for at least some specific well-defined clock mechanisms. With this idea in mind, we shall compute and compare in this chapter the rates of twelve stationary elementary electromagnetic "clocks" with the rates of the same moving "clocks." We shall base our calculations on the fundamental laws of electromagnetism and mechanics with no input from relativity theory (although we shall use the longitudinal and transverse masses, which may be regarded either as experimentally obtained quantities, or as relativistic concepts (see Section 8-4 and Ref. 9 in Chapter 8). The operation of our clocks will be based on the interaction between a field-experiencing electric point charge and different field-producing electric charge configurations.

As we shall see, some moving clocks do indeed run in agreement with the Einstein's theory,[9] but others do not.

10-2. Clocks Running in Accordance with Einstein's Special Relativity Theory

Clock #1. Consider a ring of radius a carrying a uniformly distributed charge q_1. Let the axis of the ring be the x axis, and let the center of the ring be the origin of rectangular coordinates (Fig. 10.1). The electric field on the axis of the ring is[10]

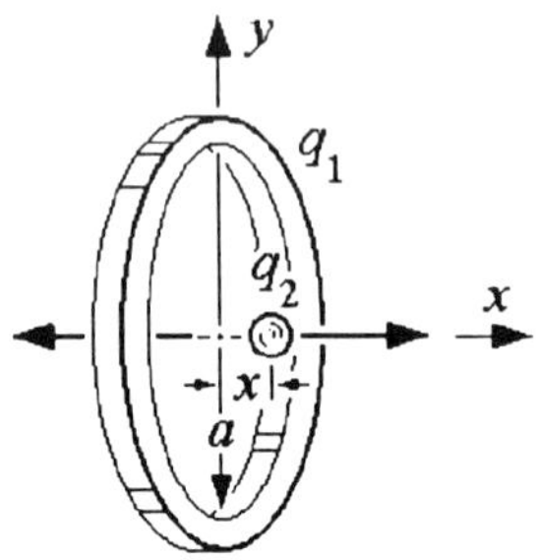

Fig. 10.1 A point charge q_2 is placed on the axis of an oppositely charged ring carrying a charge q_1. The point charge oscillates along the x axis about the center of the ring. This system can be used as a primitive clock.

$$\mathbf{E} = \frac{q_1 x}{4\pi\varepsilon_0(a^2+x^2)^{3/2}}\mathbf{i}. \qquad (10\text{-}2.1)$$

A charge q_2, whose polarity is opposite to that of q_1 and whose mass is m_0, is placed on the x axis near the center of the ring at a distance x from the center and is constrained to move only along the axis.[11] By Eq. (10-2.1), if q_2 is sufficiently close to the center, so that $x \ll a$, which we assume to be the case, the force on q_2, $F = q_2 E$, is essentially

$$\mathbf{F} = -\frac{q_1 q_2 x}{4\pi\varepsilon_0 a^3}\mathbf{i}. \qquad (10\text{-}2.2)$$

Let the ring be fixed in the laboratory. Since the force given by Eq. (10-2.2) is a linear restoring force, the ring and the charge constitute a simple harmonic oscillator, and the period of the oscillations of q_2 is

$$T = 2\pi\left(\frac{m_0}{F/x}\right)^{1/2} = 4\pi^{3/2}a^{3/2}\left(\frac{m_0\varepsilon_0}{q_1 q_2}\right)^{1/2}. \qquad (10\text{-}2.3)$$

Clearly, the ring and the charge may be considered to constitute a clock and can be used for measuring time in terms of the period of oscillations T.

Let us now assume that the same ring and the charge q_2 are located in a reference frame moving along the x axis with velocity $\mathbf{v} = v\mathbf{i}$ relative to the laboratory. By symmetry, the electric field on the axis of the ring is the same as the x component of the electric field of a moving point charge q_1 whose perpendicular distance from the axis is a. The electric field of a moving point charge is given by Heaviside's Eq. (4-1.13)[12]

$$\mathbf{E}_m = \frac{q(1-v^2/c^2)}{4\pi\varepsilon_0 r^3[1-(v^2/c^2)\sin^2\theta]^{3/2}}\mathbf{r}, \tag{10-2.4}$$

where v is the velocity of the charge, c is the velocity of light, $\mathbf{r}$ is the vector from the present position of the charge to the point of observation, and θ is the angle between $\mathbf{r}$ and $\mathbf{v}$; the subscript m is used to indicate that the field under consideration is that of the moving charge. Since $r^3 = (a^2 + x^2)^{3/2}$ and since $\sin^2\theta = a^2/(a^2 + x^2)$, we have for the field on the axis of the ring

$$\mathbf{E}_m = \frac{q_1(1-v^2/c^2)x}{4\pi\varepsilon_0(a^2+x^2)^{3/2}[1-v^2a^2/c^2(a^2+x^2)]^{3/2}}\mathbf{i}. \tag{10-2.5}$$

Assuming, as before, that $x \ll a$, we then have for the force on q_2

$$\mathbf{F}_m = -\frac{q_1 q_2 x}{4\pi\varepsilon_0 a^3(1-v^2/c^2)^{1/2}}\mathbf{i}. \tag{10-2.6}$$

Let us also assume that the velocity $\mathbf{v}$ of the moving reference frame is much larger than the maximum velocity of q_2 relative to the ring. In this case the velocity of q_2 relative to the laboratory is essentially v, and the *longitudinal mass* of q_2 is, by Eq. (8-4.9),

$$m_{\parallel} = \frac{m_0}{(1-v^2/c^2)^{3/2}}. \tag{10-2.7}$$

The period of the oscillations of q_2 is therefore

$$T_m = 2\pi\left(\frac{m_{\|}}{F_m/x}\right)^{1/2} = 2\pi\left[\frac{m_0 4\pi\varepsilon_0 a^3(1-v^2/c^2)^{1/2}}{(1-v^2/c^2)^{3/2}q_1 q_2}\right]^{1/2}$$
$$= 4\pi^{3/2}a^{3/2}\left[\frac{m_0\varepsilon_0}{(1-v^2/c^2)q_1 q_2}\right]^{1/2}, \tag{10-2.8}$$

so that

$$T_m = \frac{1}{(1-v^2/c^2)^{1/2}}T. \tag{10-2.9}$$

Thus the period of the oscillations of q_2 located in the moving reference frame, as observed from the laboratory (stationary) reference frame, is by the factor $(1 - v^2/c^2)^{-1/2}$ *longer* than the period of the oscillations of q_2 in the laboratory. Hence our clock consisting of the charged ring and the point charge runs *slower* when the clock is moving, and the rate of the moving clock is $(1 - v^2/c^2)^{-1/2} = \gamma$ times the rate of the same stationary clock.

Clock #2. Consider two point charges of the same magnitude and polarity located at the points $\pm a$ of the y axis. Let the magnitude of each charge be q_1 and let the charges be fixed in the laboratory. A point charge q_2, whose polarity is opposite to that of the first two charges and whose mass is m_0, is placed at a point x of the x axis close to the origin ($x \ll a$) (Fig. 10.2). By the same reasoning as in the case of Clock #1, the force on q_2 is

$$\mathbf{F} = -\frac{q_1 q_2 x}{2\pi\varepsilon_0 a^3}\mathbf{i}. \tag{10-2.10}$$

Therefore this system, too, is a harmonic oscillator, and the period of oscillations of q_2 is

$$T = 2\pi\left(\frac{m_0}{F/x}\right)^{1/2} = (2\pi a)^{3/2}\left(\frac{m_0\varepsilon_0}{q_1 q_2}\right)^{1/2}. \tag{10-2.11}$$

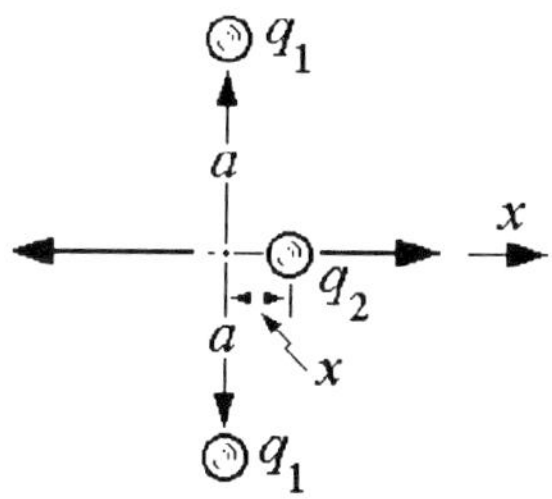

Fig. 10.2 *A point charge q_2 oscillates under the action of the two fixed point charges q_1 whose polarity is opposite to that of q_2. This system can be used as a primitive clock.*

Clearly, the three charges can be regarded as a clock for measuring time in terms of the period of oscillations T.

As in the case of Clock #1, if the three charges are placed in a reference frame moving along the x axis with velocity $\mathbf{v} = v\mathbf{i}$ relative to the laboratory, the charge q_2 will experience a force

$$\mathbf{F}_m = -\frac{q_1 q_2 x}{2\pi\varepsilon_0 a^3(1 - v^2/c^2)^{1/2}}\mathbf{i}. \tag{10-2.12}$$

Therefore q_2 will oscillate with a period

$$T_m = (2\pi a)^{3/2}\left[\frac{m_0\varepsilon_0}{(1 - v^2/c^2)q_1 q_2}\right]^{1/2}, \tag{10-2.13}$$

so that

$$T_m = \frac{1}{(1 - v^2/c^2)^{1/2}}T. \tag{10-2.14}$$

Hence our clock consisting of the three charges runs *slower* when the clock is moving, and the rate of the moving clock is $(1 - v^2/c^2)^{-1/2} = \gamma$ times the rate of the same stationary clock.

Clock #3. Consider now the same system of three charges but in a different configuration relative to the coordinate axes. Let the q_1 charges be located on the z axis at distances $\pm a$ from the origin, and let the charge q_2 be located on the y axis at a distance y close to the origin ($y \ll a$) (Fig. 10.3). The electric field at the location of q_2 is now

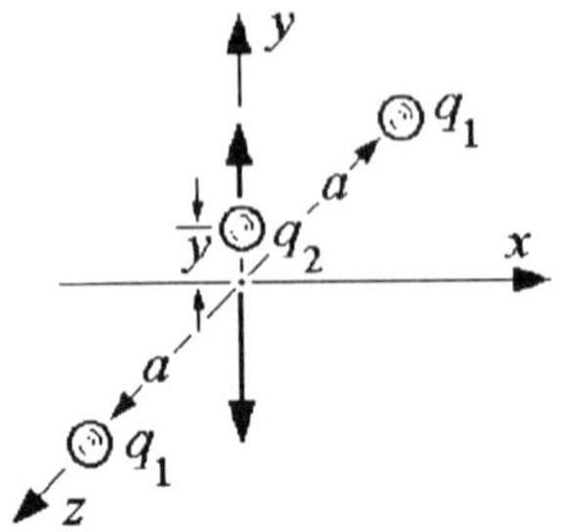

Fig. 10.3 *A point charge q_2 oscillates along the y axis under the action of the two fixed point charges q_1 whose polarity is opposite to that of q_2. This system can be used as a primitive clock.*

$$\mathbf{E} = \frac{q_1 y}{2\pi\varepsilon_0(a^2+y^2)^{3/2}}\mathbf{j}, \tag{10-2.15}$$

which, after neglecting y^2 in the denominator, becomes

$$\mathbf{E} = \frac{q_1 y}{2\pi\varepsilon_0 a^3}\mathbf{j}. \tag{10-2.16}$$

The force on q_2 is therefore

$$\mathbf{F} = -\frac{q_1 q_2 y}{2\pi\varepsilon_0 a^3}\mathbf{j}. \tag{10-2.17}$$

Except for the direction, this is the same force as that given by Eq. (10-2.10). Therefore q_2 executes a simple harmonic motion with the period given by Eq. (10-2.11) (with x replaced by y).

Let us now assume that the three charges are placed in a reference frame moving along the x axis with velocity $\mathbf{v} = v\mathbf{i}$ relative to the laboratory. In determining the force on q_2, we must now take into account that q_2 is subjected not only to the electric field but also to the magnetic field. As seen from the laboratory, the force on q_2 is therefore the Lorentz force (we assume, as before, that the velocity of q_2 is essentially v)

$$\mathbf{F}_L = q_2(\mathbf{E}_m + \mathbf{v}\times\mathbf{B}_m), \tag{10-2.18}$$

where $\mathbf{E}_m$ is the electric field, and $\mathbf{B}_m$ is the magnetic flux density field produced at the location of q_2 by the moving charges q_1.

The electric field at the location of q_2 is given by Eq. (10-2.4) with $q = q_1$, $\mathbf{r} = y\mathbf{j}$, $r = (a^2 + y^2)^{1/2}$, $\sin\theta = 1$, and with the factor 2 instead of 4 in the denominator, that is

$$\mathbf{E}_m = \frac{q_1(1 - v^2/c^2)y}{2\pi\varepsilon_0(a^2 + y^2)^{3/2}[1 - v^2/c^2]^{3/2}}\mathbf{j}, \tag{10-2.19}$$

which, after neglecting y^2, becomes

$$\mathbf{E}_m = \frac{q_1 y}{2\pi\varepsilon_0 a^3(1 - v^2/c^2)^{1/2}}\mathbf{j}. \tag{10-2.20}$$

By Eq. (3-2.6), the electric and magnetic fields of a uniformly moving charge distribution are connected by the formula

$$\mathbf{B}_m = (\mathbf{v} \times \mathbf{E}_m)/c^2. \tag{10-2.21}$$

Therefore, by Eq. (10-2.18), (10-2.20), and (10-2.21), we have for the Lorentz force acting on q_2

$$\mathbf{F}_L = -\frac{q_1 q_2 y}{2\pi\varepsilon_0 a^3(1 - v^2/c^2)^{1/2}}\left[\mathbf{j} + \frac{\mathbf{v} \times (\mathbf{v} \times \mathbf{j})}{c^2}\right], \tag{10-2.22}$$

or

$$\mathbf{F}_L = -\frac{q_1 q_2 y}{2\pi\varepsilon_0 a^3}\left(1 - \frac{v^2}{c^2}\right)^{1/2}\mathbf{j}. \tag{10-2.23}$$

Using now the *transverse mass* of q_2 [see Eq. (8-4.12)]

$$m_\perp = \frac{m_0}{(1 - v^2/c^2)^{1/2}}, \tag{10-2.24}$$

we obtain for the period of the oscillations of q_2

$$T_m = 2\pi\left(\frac{m_\perp}{F_L/y}\right)^{1/2} = (2\pi a)^{3/2}\left[\frac{m_0\varepsilon_0}{(1 - v^2/c^2)q_1 q_2}\right]^{1/2}. \tag{10-2.25}$$

Once again therefore

$$T_m = \frac{1}{(1 - v^2/c^2)^{1/2}} T, \tag{10-2.26}$$

so that our clock consisting of the three charges runs *slower* when the clock is moving, and the rate of the moving clock is $(1 - v^2/c^2)^{-1/2} = \gamma$ times the rate of the same stationary clock.

Clock #4. Consider two point charges q_1 and q_2 of the same polarity located at a distance r one from the other (Fig. 10.4). Let q_1 be fixed in the laboratory and let q_2 be free to move under the action of q_1. The force exerted by q_1 upon q_2 is

$$\mathbf{F} = \frac{q_1 q_2}{4\pi\varepsilon_0 r^3}\mathbf{r}. \tag{10-2.27}$$

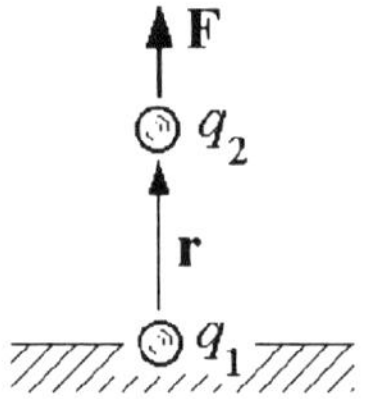

Fig. 10.4 A point charge q_2 moves under the action of the point charge q_1 whose polarity is the same as that of q_2. This system can be used as a primitive clock.

If r is sufficiently large, and if q_2 moves only a short distance, which we assume to be the case, we can ignore the variation of the force with r, so that the force can be considered essentially constant.[13] Let the mass of q_2 be m_0. The distance traveled by q_2 during a time interval Δt (as measured by the "standard clock" in the laboratory) is then

$$d = \frac{F}{2m_0}(\Delta t)^2 = \frac{q_1 q_2}{8\pi\varepsilon_0 m_0 r^2}(\Delta t)^2. \tag{10-2.28}$$

Hence we can use the two charges as a clock for measuring time intervals in terms of the distance d traveled by q_2. By Eq. (10-2.28), the formula for converting d into Δt is

$$\Delta t = \left(\frac{8\pi\varepsilon_0 m_0 r^2}{q_1 q_2} d \right)^{1/2}. \tag{10-2.29}$$

Note that the rate of our two-charge clock depends on how fast q_2 travels under the action of q_1: for a given d, the larger Δt is, the slower the rate of the clock.

Let us now assume that we have a second two-charge clock, identical with the one just described, but located in a reference frame that moves along the x axis with velocity $\mathbf{v} = v\mathbf{i}$ relative to the laboratory. Let us also assume that the line joining the two charges is perpendicular to $\mathbf{v}$, and let us assume that the velocity which q_2 acquires under the action of q_1 is much smaller than v. As seen from the laboratory, the force on q_2 is then the Lorentz force

$$\mathbf{F}_L = q_2(\mathbf{E}_m + \mathbf{v} \times \mathbf{B}_m), \tag{10-2.30}$$

where $\mathbf{E}_m$ is the electric field, and $\mathbf{B}_m$ is the magnetic flux density field produced at the location of q_2 by the moving q_1.

Since the line joining the two charges is perpendicular to $\mathbf{v}$, so that $\sin\theta = 1$ in Eq. (10-2.4), the electric field $\mathbf{E}_m$ is

$$\mathbf{E}_m = \frac{q_1}{4\pi\varepsilon_0 r^3 (1 - v^2/c^2)^{1/2}} \mathbf{r}, \tag{10-2.31}$$

and the magnetic flux density field is

$$\mathbf{B}_m = \frac{\mathbf{v} \times \mathbf{E}_m}{c^2} = \frac{q_1}{4\pi\varepsilon_0 r^3 c^2 (1 - v^2/c^2)^{1/2}} \mathbf{v} \times \mathbf{r}. \tag{10-2.32}$$

Hence the Lorentz force on q_2 is

$$\mathbf{F}_L = \frac{q_1 q_2}{4\pi\varepsilon_0 r^3 (1 - v^2/c^2)^{1/2}} \left[\mathbf{r} + \frac{\mathbf{v} \times (\mathbf{v} \times \mathbf{r})}{c^2} \right], \tag{10-2.33}$$

or

$$\mathbf{F}_L = \frac{q_1 q_2}{4\pi\varepsilon_0 r^3}\left(1 - \frac{v^2}{c^2}\right)^{1/2}\mathbf{r}. \qquad (10\text{-}2.34)$$

Using now the *transverse mass* of q_2 [see Eq. (10-2.24)], we obtain for the distance traveled by q_2 under the action of q_1

$$d_m = \frac{F_L}{2m_\perp}(\Delta t_m)^2 = \frac{q_1 q_2(1 - v^2/c^2)}{8\pi\varepsilon_0 m_0 r^2}(\Delta t_m)^2, \qquad (10\text{-}2.35)$$

where the subscripts m are used to indicate that we are now dealing with the moving two-charge clock. According to Eq. (10-2.35), the time interval needed for q_2 to travel through the distance d_m is

$$\Delta t_m = \left[\frac{8\pi\varepsilon_0 m_0 r^2}{q_1 q_2(1 - v^2/c^2)}d_m\right]^{1/2}. \qquad (10\text{-}2.36)$$

Let us now compare Δt and Δt_m corresponding to equal distances traveled by q_2 under the action of q_1 in the stationary and in the moving two-charge clock, that is, corresponding to

$$d_m = d. \qquad (10\text{-}2.37)$$

From Eqs. (10-2.29), (10-2.36), and (10-2.37) we have

$$\Delta t_m = \frac{1}{(1 - v^2/c^2)^{1/2}}\Delta t. \qquad (10\text{-}2.38)$$

Thus Δt_m is by the factor $(1 - v^2/c^2)^{-1/2}$ *longer* than Δt. Hence our moving two-charge clock runs $(1 - v^2/c^2)^{-1/2} = \gamma$ times *slower* than the identical stationary clock.

Clock #5. This clock is similar to Clock #4 just discussed, except that the fixed point charge q_1 is now replaced by a long line charge of uniform line density λ lying along the z axis of rectangular coordinates and having its midpoint at the origin (Fig. 10.5). The point charge q (no subscript is needed now) is placed on the y axis at a distance r from the origin.

Fig. 10.5 A point charge q moves under the action of the line charge λ whose polarity is the same as that of q. This system can be used as a primitive clock.

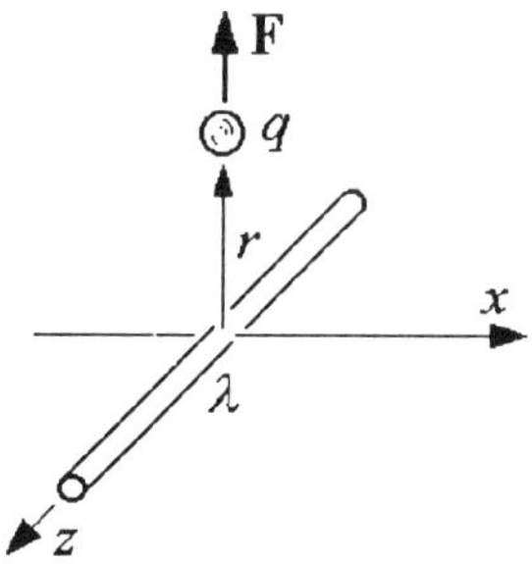

The electric field produced by the line charge at the location of q is[14]

$$\mathbf{E} = \frac{\lambda}{2\pi\varepsilon_0 r}\mathbf{j}, \tag{10-2.39}$$

and the force exerted by the line charge upon q is

$$\mathbf{F} = \frac{q\lambda}{2\pi\varepsilon_0 r}\mathbf{j}. \tag{10-2.40}$$

As before, if r is sufficiently large, and if q moves only a short distance, which we assume to be the case, we can ignore the variation of the force with r, so that the force can be considered essentially constant.[13] Let the mass of q be m_0. The distance traveled by q during a time interval Δt (as measured by the "standard clock" in the laboratory) is then

$$d = \frac{F}{2m_0}(\Delta t)^2 = \frac{q\lambda}{4\pi\varepsilon_0 m_0 r}(\Delta t)^2. \tag{10-2.41}$$

Hence we can use this line charge and the point charge as a clock for measuring time in terms of the distance d traveled by q in accordance with

$$\Delta t = \left(\frac{4\pi\varepsilon_0 m_0 r}{q\lambda}d\right)^{1/2}. \tag{10-2.42}$$

Let us now assume that we have a second clock, identical with the one just described, but located in a reference frame moving along the x axis with velocity $\mathbf{v}$ relative to the laboratory. Let us also assume that the velocity which q acquires under the action of the line charge is much smaller than v. As seen from the laboratory, the force on q is then the Lorentz force

$$\mathbf{F}_L = q(\mathbf{E}_m + \mathbf{v} \times \mathbf{B}_m), \tag{10-2.43}$$

where $\mathbf{E}_m$ is the electric field, and $\mathbf{B}_m$ is the magnetic flux density field produced at the location of q by the moving line charge.

The electric field of the moving line charge can be found by integrating Eq. (10-2.4). Since the integration is rather simple, we shall not reproduce it here and shall merely state the result:[15]

$$\mathbf{E}_m = \frac{\lambda}{2\pi\varepsilon_0 r(1 - v^2/c^2)^{1/2}}\mathbf{j}. \tag{10-2.44}$$

The Lorentz force acting on q is therefore, by Eqs. (10-2.43), (10-2.21) and (10-2.44),

$$\mathbf{F}_L = \frac{q\lambda}{2\pi\varepsilon_0 r(1 - v^2/c^2)^{1/2}}\left[\mathbf{j} + \frac{\mathbf{v} \times (\mathbf{v} \times \mathbf{j})}{c^2}\right] = \frac{q\lambda}{2\pi\varepsilon_0 r}\left(1 - \frac{v^2}{c^2}\right)^{1/2}\mathbf{j}. \tag{10-2.45}$$

Using now the *transverse mass* of q [see Eq. (10-2.24)], we obtain for the distance traveled by q under the action of the line charge

$$d_m = \frac{F_L}{2m_\perp}(\Delta t_m)^2 = \frac{q\lambda(1 - v^2/c^2)}{4\pi\varepsilon_0 m_0 r}(\Delta t_m)^2. \tag{10-2.46}$$

The time interval needed for q to travel through the distance d_m is then

$$\Delta t_m = \left[\frac{4\pi\varepsilon_0 m_0 r}{q\lambda(1 - v^2/c^2)} d_m\right]^{1/2}. \tag{10-2.47}$$

Let us now compare Δt and Δt_m corresponding to equal distances traveled by q under the action of λ in the laboratory and in the moving reference frame, that is, corresponding to

$$d_m = d. \tag{10-2.48}$$

From Eqs. (10-2.42), (10-2.47), and (10-2.48) we have

$$\Delta t_m = \frac{1}{(1 - v^2/c^2)^{1/2}} \Delta t. \tag{10-2.49}$$

Thus Δt_m is by the factor $(1 - v^2/c^2)^{-1/2}$ *longer* than Δt. Hence our moving clock consisting of the line charge and the point charge runs $(1 - v^2/c^2)^{-1/2} = \gamma$ times *slower* than the identical stationary clock.

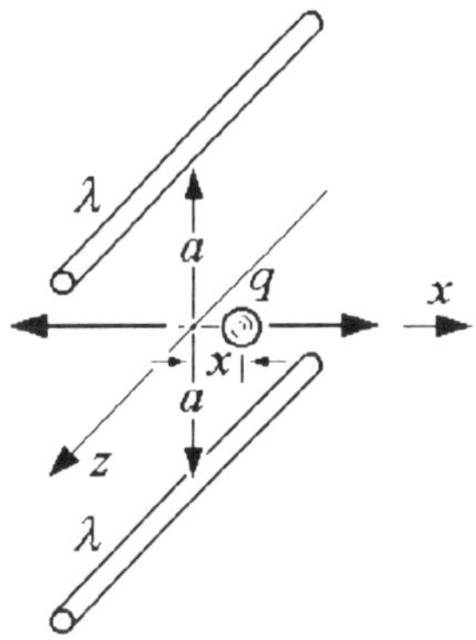

Fig. 10.6 A point charge q oscillates along the x axis under the action of two fixed line charges λ. This system can be used as a primitive clock.

Clock #6. This clock is similar to Clock #2 except that instead of the two point charges two infinitely long line charges of line charge density λ are now placed in the yz plane parallel to the z axis at distances $\pm a$ from the axis. The point charge q (there is no need for a subscript now) is again on the x axis close to the origin ($x \ll a$) (Fig. 10.6).

The electric field produced by the line charges at the location of q is[14]

$$\mathbf{E} = \frac{\lambda x}{\pi\varepsilon_0(a^2 + y^2)} \mathbf{i}, \tag{10-2.50}$$

or, since $x \ll a$,

$$\mathbf{E} = \frac{\lambda x}{\pi\varepsilon_0 a^2}\mathbf{i}. \tag{10-2.51}$$

The force experienced by q is therefore a linear restoring force

$$\mathbf{F} = -\frac{q\lambda x}{\pi\varepsilon_0 a^2}\mathbf{i} \tag{10-2.52}$$

causing q to oscillate with the period

$$T = 2\pi\left(\frac{m_0}{F/x}\right)^{1/2} = 2\pi a\left(\frac{m_0\varepsilon_0\pi}{q\lambda}\right)^{1/2}. \tag{10-2.53}$$

Let us now assume that the same field-producing line charges λ and the point charge q are located in a reference frame moving along the x axis with velocity $\mathbf{v} = v\mathbf{i}$ relative to the laboratory.

As seen from the laboratory, q now experiences an electric field which can be obtained by integrating Eq. (10-2.4) and is[15]

$$\mathbf{E}_m = \frac{\lambda(1-v^2/c^2)^{1/2}}{\pi\varepsilon_0[x^2+a^2(1-v^2/c^2)]}x\mathbf{i}, \tag{10-2.54}$$

or, since $x \ll a$,

$$\mathbf{E}_m = \frac{\lambda}{\pi\varepsilon_0 a^2(1-v^2/c^2)^{1/2}}x\mathbf{i}. \tag{10-2.55}$$

The force acting upon q is therefore

$$\mathbf{F}_m = -\frac{q\lambda}{\pi\varepsilon_0 a^2(1-v^2/c^2)^{1/2}}x\mathbf{i} \tag{10-2.56}$$

causing q to oscillate with the period [observe that we must now use the *longitudinal mass*, see Eq. (10-2.7)]

$$T_m = 2\pi\left(\frac{m_{||}}{F_m/x}\right)^{1/2} = 2\pi a\left(\frac{m_0\varepsilon_0\pi}{(1-v^2/c^2)q\lambda}\right)^{1/2}. \tag{10-2.57}$$

Thus, by Eqs. (10-2.53) and (10-2.57), the relation between the periods of our moving Clock #6 and of the same stationary clock is

$$T_m = \frac{1}{(1 - v^2/c^2)^{1/2}} T. \tag{10-2.58}$$

Once again T_m is by the factor $(1 - v^2/c^2)^{-1/2}$ *longer* than T. Hence our moving Clock #6 runs $(1 - v^2/c^2)^{-1/2} = \gamma$ times *slower* than the identical stationary clock.

10-3. Clocks that do not Run in Accordance with Einstein's Special Relativity Theory

Clock #7. This clock is the same as Clock #3 except that the field-producing charges q_1 are now placed along the x axis at the points $\pm a$ of the axis. The point charge q_2 is again on the y axis close to the origin ($y \ll a$) (Fig. 10.7).

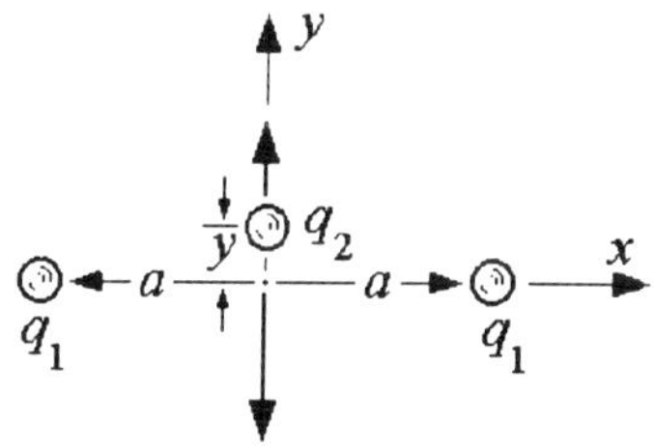

Fig. 10.7 A point charge q_2 oscillates along the y axis under the action of two fixed point charges q_1. This system can be used as a primitive clock.

Clearly, the period of the oscillations of q_2 in the laboratory frame is the same as for Clock #3, that is

$$T = 2\pi\left(\frac{m_0}{F/y}\right)^{1/2} = (2\pi a)^{3/2}\left(\frac{m_0\varepsilon_0}{q_1 q_2}\right)^{1/2}. \tag{10-3.1}$$

Let us now assume that the same field-producing charges q_1 and the charge q_2 are located in a reference frame moving along the x axis with velocity $\mathbf{v} = v\mathbf{i}$ relative to the laboratory.

As seen from the laboratory, q_2 now experiences an electric field and a magnetic field. The electric field is given by Eq. (10-2.4) with $q = q_1$, $\mathbf{r} = y\mathbf{j}$, $r = (a^2 + y^2)^{1/2}$, $\sin^2\theta = y^2/(a^2 + y^2)$, and the factor 2 instead of 4 in the denominator, that is

$$\mathbf{E}_m = \frac{q_1(1 - v^2/c^2)y}{2\pi\varepsilon_0(a^2 + y^2)^{3/2}\{1 - (v^2/c^2)[y^2/(a^2 + y^2)]\}^{3/2}}\mathbf{j}, \quad (10\text{-}3.2)$$

or, remembering that $y \ll a$ and neglecting y^2 and $y^2/(a^2 + y^2)$,

$$\mathbf{E}_m = \frac{q_1(1 - v^2/c^2)y}{2\pi\varepsilon_0 a^3}\mathbf{j}. \quad (10\text{-}3.3)$$

The magnetic field, according to Eqs. (10-2.21) and (10-3.3), is

$$\mathbf{B}_m = \frac{q_1(1 - v^2/c^2)y}{2\pi\varepsilon_0 c^2 a^3}\mathbf{v} \times \mathbf{j}. \quad (10\text{-}3.4)$$

The Lorentz force on q_2 is therefore

$$\mathbf{F}_L = -\frac{q_1 q_2(1 - v^2/c^2)y}{2\pi\varepsilon_0 a^3}\left[\mathbf{j} + \frac{\mathbf{v} \times (\mathbf{v} \times \mathbf{j})}{c^2}\right], \quad (10\text{-}3.5)$$

or

$$\mathbf{F}_L = -\frac{q_1 q_2(1 - v^2/c^2)^2}{2\pi\varepsilon_0 a^3}y\mathbf{j}. \quad (10\text{-}3.6)$$

Using now the *transverse mass* of q_2 [see Eq. (8-4.12)]

$$m_\perp = \frac{m_0}{(1 - v^2/c^2)^{1/2}}, \quad (10\text{-}3.7)$$

we obtain for the period of the oscillations of q_2

$$T_m = 2\pi\left(\frac{m_\perp}{F_L/y}\right)^{1/2} = (2\pi a)^{3/2}\left[\frac{m_0\varepsilon_0}{(1 - v^2/c^2)^{5/2}q_1 q_2}\right]^{1/2}. \quad (10\text{-}3.8)$$

Thus the relation between the periods of this moving clock and of the same stationary clock is

$$T_m = \frac{1}{(1 - v^2/c^2)^{5/4}} T, \tag{10-3.9}$$

so that although our clock consisting of the three charges runs *slower* when the clock is moving, the rate of the moving clock is $(1 - v^2/c^2)^{-5/4} = \gamma^{5/2}$ times the rate of the same stationary clock.

Fig. 10.8 A point charge q moves under the action of the fixed line charge λ. This system can be used as a primitive clock.

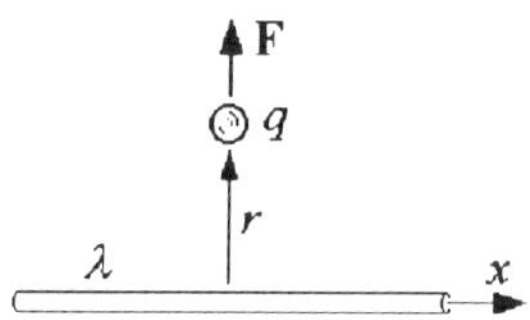

Clock #8. This clock is similar to Clock #5, except that the line charge is now placed along the x axis (the midpoint of the line charge is, as before, at the origin; see Fig. 10.8. If the clock is stationary, the force on q is the same as in the case of Clock #5 and is given by Eq. (10-2.40). Therefore the distance traveled by q during a time interval Δt (as measured by the "standard clock" in the laboratory) is also the same as for Clock #5 and is given by Eq. (10-2.41). Consequently, the time interval needed for q to travel through the distance d is also the same as for Clock #5 given by Eq. (10-2.42), that is

$$\Delta t = \left(\frac{4\pi\varepsilon_0 m_0 r}{q\lambda} d\right)^{1/2}. \tag{10-3.10}$$

Let us now assume that a second line-charge point-charge clock, identical with the one just described, is located in a reference frame that is moving along the x axis with velocity $\mathbf{v} = v\mathbf{i}$ relative to the laboratory.

By Eq. (5-3.13) or Eq. (9-2.4) the electric field produced by the moving infinitely long line charge is the same as the field produced by the stationary charge, so that [see Eq. (10-2.39)][16]

$$\mathbf{E}_m = \frac{\lambda}{2\pi\varepsilon_0 r}\mathbf{j}. \tag{10-3.11}$$

The magnetic field produced by the line charge is then, by Eq. (10-2.21)

$$\mathbf{B}_m = \frac{\lambda}{2\pi\varepsilon_0 c^2 r}\mathbf{v}\times\mathbf{j}, \tag{10-3.12}$$

and the Lorentz force acting on q is

$$\mathbf{F}_L = \frac{q\lambda}{2\pi\varepsilon_0 r}\left[\mathbf{j} + \frac{\mathbf{v}\times(\mathbf{v}\times\mathbf{j})}{c^2}\right] = \frac{q\lambda}{2\pi\varepsilon_0 r}\left(1 - \frac{v^2}{c^2}\right)\mathbf{j}. \tag{10-3.13}$$

Using now the *transverse mass* of q [see Eq. (10-3.7)], we obtain for the distance traveled by q under the action of the line charge

$$d_m = \frac{F_L}{2m_\perp}(\Delta t_m)^2 = \frac{q\lambda(1 - v^2/c^2)^{3/2}}{4\pi\varepsilon_0 m_0 r}(\Delta t_m)^2. \tag{10-3.14}$$

The time interval needed for q to travel through the distance d_m is then

$$\Delta t_m = \left(\frac{4\pi\varepsilon_0 m_0 r}{q\lambda(1 - v^2/c^2)^{3/2}}d_m\right)^{1/2}. \tag{10-3.15}$$

Let us now compare Δt and Δt_m corresponding to equal distances traveled by q under the action of λ in the laboratory and in the moving reference frame, that is, corresponding to

$$d_m = d. \tag{10-3.16}$$

From Eqs. (10-3.10), (10-3.15), and (10-3.16) we have

$$\Delta t_m = \frac{1}{(1 - v^2/c^2)^{3/4}}\Delta t. \tag{10-3.17}$$

Hence although our clock consisting of a line charge and a point charge runs *slower* when the clock is moving, the rate of the moving clock is $(1 - v^2/c^2)^{-3/4} = \gamma^{3/2}$ times the rate of the same stationary clock.

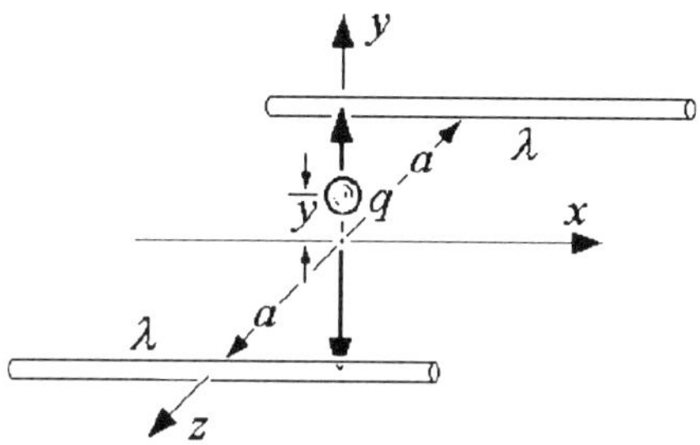

Fig. 10.9 A point charge q oscillates along the y axis under the action of two line charges λ. This system can be used as a primitive clock.

Clock #9. This clock is similar to Clock #3 except that instead of the two point charges two infinitely long line charges of line charge density λ are now placed in the xz plane parallel to the x axis at distances $\pm a$ from the axis. The point charge q (there is no need for the subscript now) is again on the y axis close to the origin ($y \ll a$) (Fig. 10.9).

The electric field produced by the line charges at the location of q is[14]

$$\mathbf{E} = \frac{\lambda y}{\pi \varepsilon_0 (a^2 + y^2)} \mathbf{j}, \qquad (10\text{-}3.18)$$

or, since $y \ll a$,

$$\mathbf{E} = \frac{\lambda y}{\pi \varepsilon_0 a^2} \mathbf{j}. \qquad (10\text{-}3.19)$$

The force experienced by q is therefore a linear restoring force

$$\mathbf{F} = -\frac{q \lambda y}{\pi \varepsilon_0 a^2} \mathbf{j} \qquad (10\text{-}3.20)$$

causing q to oscillate with the period

$$T = 2\pi\left(\frac{m_0}{F/y}\right)^{1/2} = 2\pi a\left(\frac{m_0\varepsilon_0\pi}{q\lambda}\right)^{1/2}. \qquad (10\text{-}3.21)$$

Let us now assume that the same field-producing line charges λ and the point charge q are located in a reference frame moving along the x axis with velocity $\mathbf{v} = v\mathbf{i}$ relative to the laboratory.

As seen from the laboratory, q now experiences an electric field and a magnetic field. The electric field, by Eq. (5-3.13) or Eq. (9-2.4), is the same as the field produced by the stationary line charges, that is

$$\mathbf{E}_m = \frac{\lambda y}{\pi\varepsilon_0 a^2}\mathbf{j}. \qquad (10\text{-}3.22)$$

The magnetic field, according to Eq. (9-2.10), is

$$\mathbf{B}_m = \frac{\lambda y}{\pi\varepsilon_0 a^2 c^2}\mathbf{v}\times\mathbf{j}. \qquad (10\text{-}3.23)$$

The Lorentz force on q is therefore

$$\mathbf{F}_L = -\frac{q\lambda y}{\pi\varepsilon_0 a^2}\left[\mathbf{j} + \frac{\mathbf{v}\times(\mathbf{v}\times\mathbf{j})}{c^2}\right], \qquad (10\text{-}3.24)$$

or

$$\mathbf{F}_L = -\frac{q\lambda(1 - v^2/c^2)}{\pi\varepsilon_0 a^2}y\mathbf{j}. \qquad (10\text{-}3.25)$$

Using now the *transverse mass* of q [see Eq. (10-3.7)], we obtain for the period of the oscillations of q

$$T_m = 2\pi\left(\frac{m_\perp}{F_L/y}\right)^{1/2} = 2\pi a\left[\frac{m_0\varepsilon_0\pi}{(1 - v^2/c^2)^{3/2}q\lambda}\right]^{1/2}. \qquad (10\text{-}3.26)$$

Thus the relation between the periods of this moving clock and of the same stationary clock is

$$T_m = \frac{1}{(1 - v^2/c^2)^{3/4}} T, \tag{10-3.27}$$

so that although our Clock #9 runs *slower* when the clock is moving, the rate of the moving clock is $(1 - v^2/c^2)^{-3/4} = \gamma^{3/2}$ times the rate of the same stationary clock.

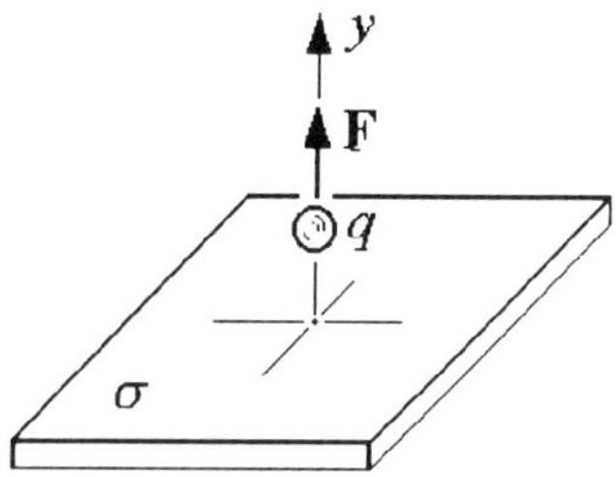

Fig. 10.10 A point charge q moves along the y axis under the action of a charged plate carrying a surface charge σ. This system can be used as a primitive clock.

Clock #10. Consider a large uniformly charged plate of surface charge density σ. Let the plate be in the xz plane with its center at the origin. A point charge q of the same polarity as the plate is placed on the y axis not far from the plate (Fig. 10.10). Let the plate be fixed in the laboratory and let q be free to move under the action of σ. The electric field produced by σ at the location of q is

$$\mathbf{E} = \frac{\sigma}{2\varepsilon_0}\mathbf{j}. \tag{10-3.28}$$

The force exerted on q is therefore

$$\mathbf{F} = \frac{\sigma q}{2\varepsilon_0}\mathbf{j}. \tag{10-3.29}$$

Let the mass of q be m_0. The distance traveled by q during a time interval Δt (as measured by the "standard clock" in the laboratory) is then

$$d = \frac{F}{2m_0}(\Delta t)^2 = \frac{\sigma q}{4\varepsilon_0 m_0}(\Delta t)^2. \tag{10-3.30}$$

Hence we can use the plate and the charge as a clock for measuring time intervals in terms of the distance d traveled by q. The formula for converting d into Δt is, by Eq. (10-3.30),

$$\Delta t = \left(\frac{4\varepsilon_0 m_0}{\sigma q} d\right)^{1/2}. \tag{10-3.31}$$

Let us now assume that we have a second clock, identical with the one just described, but located in a reference frame that moves along the x axis with velocity $\mathbf{v} = v\mathbf{i}$ relative to the laboratory. Let us also assume that the velocity which q acquires under the action of σ is much smaller than v. As seen from the laboratory, the force on q is then the Lorentz force

$$\mathbf{F}_L = q(\mathbf{E}_m + \mathbf{v} \times \mathbf{B}_m), \tag{10-3.32}$$

where $\mathbf{E}_m$ is the electric field, and $\mathbf{B}_m$ is the magnetic flux density field produced at the location of q by the moving plate.

According to Section 9-2, the electric field $\mathbf{E}_m$ produced by the moving plate is the same as that of the stationary plate, that is[17]

$$\mathbf{E}_m = \frac{\sigma}{2\varepsilon_0}\mathbf{j}. \tag{10-3.33}$$

The Lorentz force acting on q is then

$$\mathbf{F}_L = \frac{\sigma q}{2\varepsilon_0}\left[\mathbf{j} + \frac{\mathbf{v} \times (\mathbf{v} \times \mathbf{j})}{c^2}\right] = \frac{\sigma q}{2\varepsilon_0}\left(1 - \frac{v^2}{c^2}\right)\mathbf{j}. \tag{10-3.34}$$

Using now the *transverse mass* of q [see Eq. (10-3.7)], we obtain for the distance traveled by q under the action of the plate

$$d_m = \frac{F_L}{2m_\perp}(\Delta t_m)^2 = \frac{q\lambda(1 - v^2/c^2)^{3/2}}{4\pi\varepsilon_0 m_0 r}(\Delta t_m)^2. \tag{10-3.35}$$

According to Eq. (10-3.35), the time interval needed for q to travel through the distance d_m is

$$\Delta t_m = \left(\frac{4\varepsilon_0 m_0}{\sigma q (1 - v^2/c^2)^{3/2}} d_m \right)^{1/2}. \tag{10-3.36}$$

Let us now compare Δt and Δt_m corresponding to equal distances traveled by q under the action of the charged plate in the stationary and in the moving clock, that is, corresponding to

$$d_m = d. \tag{10-3.37}$$

From Eqs. (10-3.31), (10-3.36), and (10-3.37) we have

$$\Delta t_m = \frac{1}{(1 - v^2/c^2)^{3/4}} \Delta t. \tag{10-3.38}$$

Hence although our clock consisting of a charged plate and the point charge runs *slower* when the clock is moving, the rate of the moving clock is $(1 - v^2/c^2)^{-3/4} = \gamma^{3/2}$ times the rate of the same stationary clock.

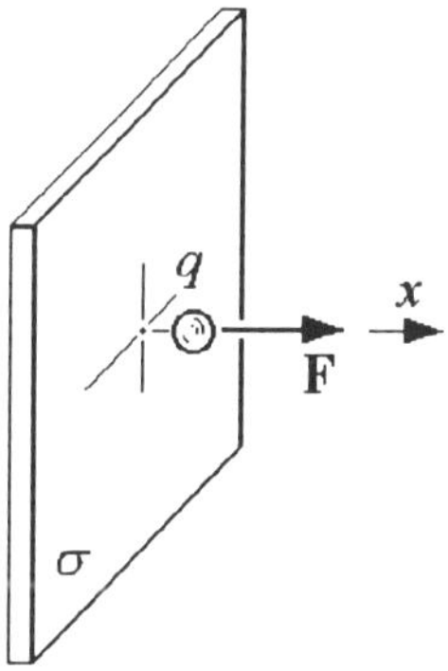

Fig. 10.11 A point charge q moves along the x axis under the action of a charged plate carrying a surface charge σ. This system can be used as a primitive clock.

Clock #11. This clock is the same as the one just discussed, except that the plate is now in the *yz* plane and the point charge q is on the x axis. The clock at rest in the laboratory behaves exactly like Clock #10, and therefore Eq. (10-3.31) applies also to the present clock, so that

$$\Delta t = \left(\frac{4\varepsilon_0 m_0}{\sigma q} d\right)^{1/2}. \tag{10-3.39}$$

The electric field produced at the location of q by the moving plate, as seen from the laboratory, can be obtained by integrating Eq. (10-2.4). However, this time we shall obtain the field by making use of the method of corresponding states (see Section 7-6). Taking into account that the charge and the surface area of the plate are invariant under Lorentz transformations and using Eq. (7-1.5), we find that the electric field of the moving plate at the location of q is the same as that of the stationary plate.[18] Furthermore, since the electric field is parallel to the velocity of the plate, the plate produces no magnetic field at the location of q. Hence the force exerted by the moving plate on q is

$$\mathbf{F}_m = \frac{\sigma q}{2\varepsilon_0}\mathbf{i}. \tag{10-3.40}$$

Using the *longitudinal mass* [see Eq. (8-4.9)],

$$m_{\|} = \frac{m_0}{(1 - v^2/c^2)^{3/2}}, \tag{10-3.41}$$

we then find that the distance traveled by q under the action of the moving plate, as seen from the laboratory, is

$$d_m = \frac{F_m}{2m_{\|}}(\Delta t_m)^2 = \frac{\sigma q(1 - v^2/c^2)^{3/2}}{4\varepsilon_0 m_0}(\Delta t_m)^2, \tag{10-3.42}$$

so that

$$\Delta t_m = \left[\frac{4\varepsilon_0 m_0}{\sigma q(1 - v^2/c^2)^{3/2}} d_m\right]^{1/2}. \tag{10-3.43}$$

Hence although our clock consisting of a charged plate and the point charge runs *slower* when the clock is moving, the rate of the moving clock is $(1 - v^2/c^2)^{-3/4} = \gamma^{3/2}$ times the rate of the same stationary clock.

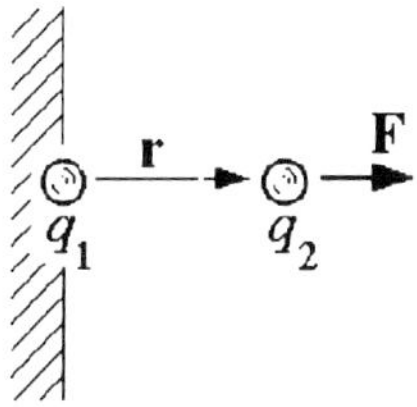

Fig. 10.12 A point charge q_2 moves along the x axis under the action of the point charge q_1. This system can be used as a primitive clock.

Clock #12. This clock is the same as Clock #4, except that the line joining the two charges is parallel to the x axis (Fig. 10.12).

In the laboratory, the clock functions exactly as Clock #4, so that the time interval given by Eq. (10-2.29)

$$\Delta t = \left(\frac{8\pi\varepsilon_0 m_0 r^2}{q_1 q_2} d\right)^{1/2} \qquad (10\text{-}3.44)$$

applies also to the present clock.

When the clock moves along the x axis with velocity $\mathbf{v} = v\mathbf{i}$ relative to the laboratory, the electric field due to charge q_1 becomes, according to Eq. (10-2.4),

$$\mathbf{E}_m = \frac{q_1(1 - v^2/c^2)}{4\pi\varepsilon_0 r^3}\mathbf{r}, \qquad (10\text{-}3.45)$$

and the force on q_2 becomes

$$\mathbf{F}_m = \frac{q_1 q_2(1 - v^2/c^2)}{4\pi\varepsilon_0 r^3}\mathbf{r}. \qquad (10\text{-}3.46)$$

Using the *longitudinal mass* [see Eq. (10-3.41)], we find that the distance traveled by q_2 under the action of q_1 is now

$$d_m = \frac{F_m}{2m_{\|}}(\Delta t_m)^2 = \frac{q_1 q_2(1 - v^2/c^2)^{5/2}}{8\pi\varepsilon_0 m_0 r^2}(\Delta t_m)^2, \qquad (10\text{-}3.47)$$

and the time interval needed for q_2 to travel the distance d_m is

$$\Delta t_m = \left[\frac{8\pi\varepsilon_0 m_0 r^2}{q_1 q_2 (1 - v^2/c^2)^{5/2}} d_m \right]^{1/2} . \qquad (10\text{-}3.48)$$

Let us now compare Δt and Δt_m corresponding to equal distances traveled by q_2 under the action of q_1 in the stationary and in the moving two-charge clock, that is, corresponding to

$$d_m = d. \qquad (10\text{-}3.49)$$

From Eqs. (10-3.44), (10-3.48), and (10-3.49) we obtain

$$\Delta t_m = \frac{1}{(1 - v^2/c^2)^{5/4}} \Delta t . \qquad (10\text{-}3.50)$$

Thus Δt_m is by the factor $(1 - v^2/c^2)^{-5/4}$ *longer* than Δt. Hence our moving two-charge clock runs $(1 - v^2/c^2)^{-5/4} = \gamma^{5/2}$ times *slower* than the identical stationary clock.

10-4. Reconciling the Theory with Experimental Data

As we have seen, the primitive electromagnetic clocks discussed above run slow when the clocks move, but their rate depends on the type of the clock and even on the orientation of the clock relative to the direction of motion. Thus, contrary to Einstein's conception, the slowing down of the moving clocks is a dynamic rather than a kinematic effect, and the slowing down is not, in general, proportional to γ. Therefore, if "time" is that which is measured by physical devices (clocks), there is no such thing as time dilation depending solely on the velocity of the clocks and being the same for all the clocks moving with the same velocity.

In fact, our calculations show that the slowing down of the clocks is not really a relativistic effect at all. The calculations that we used were based on the laws of classical electromagnetism (classical field equations and Lorentz force law) with no input from relativity theory except for the longitudinal and transverse masses (which just as well may be considered to be experimentally obtained relations).[19] However, as one can easily see, even if we used the ordinary classical expression for the mass, we would still find that moving clocks run slow. Therefore relativity theory at best improves the accuracy of the calculations but does not affect the qualitative physical essence of our results.

Furthermore, relativity theory, which, as we have seen, is derived from, and is based upon, the laws of electromagnetism, does not provide us with any information on the rate of processes other than the electromagnetic ones. In particular, it does not provide us with any information on the rate of *biological* effects, such as aging. Therefore the widely popularized idea, allegedly supported by relativity theory, that space travelers moving with a velocity close to the velocity of light age slower than their Earthbound twins is no more than an attractive hypothesis having no adequate scientific foundation.[20] Actually, as far as space travel is concerned, it is very likely that interstellar magnetic fields and other external factors will have a much stronger effect on the rate of the clocks and on the condition of space travelers than any kinematic relativistic effects.

But what about experiments[7] that are interpreted as proofs of the reality of time dilation? The only thing that these experiments really prove is that the rate of certain physical phenomena is slower in systems moving at very high speeds, which, as we have just seen, need not be regarded as a relativistic effect. Therefore it is more prudent to interpret these experiments as indicating the existence of certain velocity-dependent interactions in the systems under consideration similar to the electromagnetic interactions that made the clocks discussed in this chapter run slower when in motion. More experiments and greater accuracy are definitely needed in order to elucidate the nature of these interactions and the numerical factor (or factors) by which time-dependent phenomena in moving systems differ from the same phenomena in stationary systems.

Finally, it is important to note that we have developed in this book all the essential elements of the theory of relativity as a direct mathematical and logical extension of classical electromagnetic theory without ever using the concepts of clock synchronization,[21] Lorentz contraction, and time dilation. Therefore, although clock synchronization, Lorentz contraction, and time dilation are indispensable elements in Einstein's approach to the development of the theory of relativity,[22] they cannot be considered to constitute elements of the theory itself.

References and Remarks for Chapter 10

1. J. Larmor, "On a dynamical theory of the electric and luminiferous medium, Pt III, Relations on the Material Media," Philos. Trans. R. Soc. Lond. **190A**, 205 (1897). For a discussion see C. Kittel, "Larmor and the Prehistory of the Lorentz Transformations," Am. J. Phys. **42**, 726-729 (1974).
2. A. Einstein, "Zur Elektrodynamik bewegter Körper," Ann. Phys. **17**, 891-921 (1905).
3. A. Einstein, "Die Relativitätstheorie," Nat. Ges. Zürich. Viers. **56**, 1-14 (1911).
4. A. Einstein, "Die Relativitätstheorie," in *Kultur der Gegenwart: Physik*, 2nd ed., (Teubner, Leipzig, 1925) pp. 783-797.
5. See Arthur I. Miller, *Albert Einstein's Special Theory of Relativity* (Addison-Wesley, Reading, Massachusetts, 1981) pp. 257-264.
6. See, for example, L. Brillouin, *Relativity Reexamined* (Academic Press, New York, 1970) pp. 65-73; C. A. Zapffe, *A Reminder on* $E = mc^2$ (CAZlab, Baltimore, 1982) pp. 19-21, 79-90; Petr Beckmann, *Einstein Plus Two* (The Golem Press, Boulder, 1987) pp. 77-81; Howard C. Hayden, "Yes, moving clocks run slowly, but is time dilated?", Galilean Electrodynamics **2**, 63-66 (1991).

7. B. Rossi and D. B. Hall, "Rate of Decay of Mesotrons with Momentum," Phys. Rev. **59**, 223-228 (1941); D. H. Fisch and J. H. Smith, "Measurement of the relativistic time dilation using mu-mesons," Am. J. Phys. **31**, 342-355 (1963); J. Bailey, K. Borer, F. Combley, H. Drumm, F. Krienen, F. Lange, E. Picasso, W. von Ruden, F. J. M. Farley, J. H. Field, W. Flegel, P. M. Hattersley, "Measurements of relativistic time dilation for positive and negative muons in circular orbit," Nature **268**, 301-305 (1977).
8. For an elementary discussion see, for example, A. P. French, *Special Relativity* (Norton, New York, 1968) pp. 97-104.
9. See also Oleg D. Jefimenko, "Direct calculation of time dilation," Am. J. Phys. **64**, 812-814 (1996).
10. See, for example, Oleg D. Jefimenko, *Electricity and Magnetism*, 2nd ed., (Electret Scientific, Star City, 1989) pp. 99-100.
11. The charge must be constrained to stay on the axis because otherwise it is unstable with respect to a lateral displacement. The moving charges in all the clocks discussed later in this chapter are assumed to be similarly constrained.
12. Note that this field as well as all the electric and magnetic fields produced by moving charge distributions discussed in this chapter can be obtained from the corresponding electric fields of the same stationary charge distributions by means of the method of corresponding states (see Chapter 7, Section 6) and Lorentz-Einstein transformation. We are using nonrelativistic methods for obtaining the fields in order to emphasize the nonrelativistic electromagnetic nature of the obtained results.
13. This approximation has no effect on the conclusions reached in the chapter, because, as we shall presently see, the dependence of the force on r is exactly the same for the charges in the laboratory and for the charges in the moving reference frame. However, neglecting the dependence of the force on r simplifies the calculations and eliminates irrelevant algebraic manipulations.
14. See, for example, Ref. 10, pp. 89-90 and 98-99.
15. The electric field of an infinitely long line charge moving at right angles to its length was first derived by Oliver Heaviside in "On the Electromagnetic Effects due to the Motion of Electricity

Through a Dielectric," Phil. Mag. **27**, 324-339 (1889). Except for notation, it is the same as our Eq. (10-2.44). For a relativistic derivation see Example 7-6.2. See also Oliver Heaviside, *Electrical Papers* (MacMillan, London, 1894) Vol. II, pp. 504-518.

16. This equation was first derived by Heaviside in 1888, see Ref. 15. It can also be obtained from retarded electric potential or field integrals, see Ref. 12 in Chapter 9. For a relativistic derivation see Example 7-6.3.

17. This result was obtained by Heaviside in 1888. See Ref. 15.

18. Heaviside obtained this result in 1888 by integrating his equations for the moving point charge. See Ref. 15.

19. The concept is unquestionably pre-relativistic. See Ref. 9 in Chapter 8.

20. The colossal fame of Einstein's special relativity theory is to a large extent due to the various and numerous popularizations capitalizing on Einstein's ideas of the relativistic length contraction and time dilation which he enunciated in his 1905 article (see Ref. 2). Of the many popular "explanations" and "expositions" of Einstein's special relativity theory one of the best known is by G. Gamov, the creator of Mr. Tompkins and his adventures in the relativistic "Wonderland." See, for example, G. Gamov, *Mr Tompkins in Paperback* (Cambridge University Press, New York, 1973) (numerous editions). Unfortunately, such popularizations concentrate not on the essence and useful applications of the theory of relativity but on the sensational and paradoxical aspects of Einstein's articles, thus giving the readers a totally distorted view of the real significance and utility of the *true* theory of relativity.

21. Clock synchronization is the starting point of Einstein's 1905 article (see Ref. 2) and is customarily included in all conventional presentations of relativity theory. However, Einstein's procedure of synchronization is essentially a "Gedankenexperiment" and can hardly be actually implemented. Moreover, as far as the "clock paradox" and space travel are concerned, space travelers will most likely synchronize and adjust their clocks so that the clocks will run exactly as do Earthbound clocks regardless of any effects affecting the rate of the moving clocks.

22. See, for example, Ref. 8.

11

GRAVITATION AND COVARIANCE

Recent advances in the theory of time-dependent Newtonian gravitational fields provide the foundation for a new approach to the study of gravitation and to the investigation of the connection between gravitation and other physical phenomena and effects. The basic equations representing time-dependent gravitational fields and interactions are very similar to the basic equations representing time-dependent electromagnetic fields and interactions, and most electromagnetic equations, including Maxwell's equations and retarded field equations, have their gravitational counterparts. In this chapter we shall explore the analogy between electromagnetism and gravitation and, on the basis of this analogy, shall develop a relativistic theory of gravitation analogous to relativistic electrodynamics and incorporating relativistic mechanics. Then we shall briefly discuss the so-called "covariant formulation" of electromagnetic and gravitational equations.

11-1. Analogy of Electromagnetism with Gravitation

According to the theory of time-dependent gravitational fields,[1] gravitational forces are associated with two fields: the ordinary Newtonian gravitational field **g** and the *cogravitational* field **K** (also known as the *gravimagnetic* field or *Heaviside's* field[2]). Just like the electric field, the gravitational field **g** acts on stationary as well as on moving bodies, whereas the cogravitational field acts only on moving bodies. The two fields are assumed to propagate with a finite velocity c. The value of this velocity is as yet unknown. However, it is generally assumed that it is the same as that of the velocity of light. A summary of the basic gravitational equations for time-dependent gravitational and cogravitational fields is presented below.[3] The equations included in this summary are separated in three categories:

(1) *Basic definition equations for gravitational fields*

Gravitational field **g**

$$\mathbf{g} = \mathbf{F}/m, \tag{11-1.1}$$

Cogravitational field **K**

$$\mathbf{F} = m(\mathbf{u}\times\mathbf{K}), \tag{11-1.2}$$

Mass density ρ

$$\rho = dm/dV, \tag{11-1.3}$$

Mass current density **J**

$$\mathbf{J} = \rho\mathbf{u}. \tag{11-1.4}$$

(2) *Basic differential equations for gravitational fields*

$$\nabla\cdot\mathbf{g} = -4\pi G\rho, \tag{11-1.5}$$

$$\nabla\cdot\mathbf{K} = 0, \tag{11-1.6}$$

$$\nabla\times\mathbf{g} = -\frac{\partial\mathbf{K}}{\partial t}, \tag{11-1.7}$$

$$\nabla\times\mathbf{K} = -\frac{4\pi G}{c^2}\mathbf{J} + \frac{1}{c^2}\frac{\partial\mathbf{g}}{\partial t}. \tag{11-1.8}$$

(3) *Basic causal equations for gravitational fields*

$$\mathbf{g} = G\int \frac{[\nabla'\rho + (1/c^2)(\partial\mathbf{J}/\partial t)]}{r}\,dV', \tag{11-1.9}$$

$$\mathbf{K} = -\frac{G}{c^2}\int \frac{[\nabla' \times \mathbf{J}]}{r}\,dV', \tag{11-1.10}$$

$$\mathbf{g} = -G\int\left\{\frac{[\rho]}{r^2} + \frac{1}{rc}\frac{\partial[\rho]}{\partial t}\right\}\mathbf{r}_u\,dV' + \frac{G}{c^2}\int\frac{1}{r}\left[\frac{\partial\mathbf{J}}{\partial t}\right]dV', \tag{11-1.11}$$

$$\mathbf{K} = -\frac{G}{c^2}\int\left\{\frac{[\mathbf{J}]}{r^2} + \frac{1}{rc}\frac{\partial[\mathbf{J}]}{\partial t}\right\}\times\mathbf{r}_u\,dV'. \tag{11-1.12}$$

Observe that for time-independent systems Eq. (11-1.11) reduces to the ordinary Newtonian gravitational field.

Let us now list the basic electromagnetic equations for fields in a vacuum. Arranging them in categories similar to those used for gravitational equations, we have:

(1) *Basic definitions*

Electric field **E**

$$\mathbf{E} = \mathbf{F}/q, \tag{11-1.13}$$

Magnetic flux density field **B**

$$\mathbf{F} = q(\mathbf{u}\times\mathbf{B}), \tag{11-1.14}$$

Electric charge density ρ

$$\rho = dq/dV, \tag{11-1.15}$$

Electric convection current **J**

$$\mathbf{J} = \rho\mathbf{u}. \tag{11-1.16}$$

(2) *Maxwell's equations for electromagnetic fields in a vacuum*

$$\nabla \cdot \mathbf{E} = \rho/\varepsilon_0, \tag{11-1.17}$$

$$\nabla \cdot \mathbf{B} = 0, \tag{11-1.18}$$

$$\nabla \times \mathbf{E} = -\frac{\partial \mathbf{B}}{\partial t}, \tag{11-1.19}$$

$$\nabla \times \mathbf{B} = \mu_0 \mathbf{J} + \frac{1}{c^2}\frac{\partial \mathbf{E}}{\partial t}. \tag{11-1.20}$$

(3) *Basic causal electromagnetic equations*

$$\mathbf{E} = -\frac{1}{4\pi\varepsilon_0}\int \frac{[\nabla'\rho + (1/c^2)(\partial \mathbf{J}/\partial t)]}{r}dV', \tag{11-1.21}$$

and

$$\mathbf{B} = \frac{\mu_0}{4\pi}\int \frac{[\nabla' \times \mathbf{J}]}{r}dV'. \tag{11-1.22}$$

If we compare the gravitational equations listed above with the electromagnetic equations, we find that to each gravitational equation there corresponds an electromagnetic equation. The corresponding equations are identical except for the symbols and constants occurring in them. The relations between the corresponding symbols and constants are shown in Table 11-1.

It is clear that all equations derivable from the basic electromagnetic equations listed above have their gravitational counterparts, and that various gravitational equations can be obtained from the corresponding electromagnetic equations by simply replacing the electromagnetic symbols and constants by the corresponding gravitational symbols and constants in accordance with Table 11-1.

Table 11-1

Corresponding Electromagnetic and Gravitational Symbols and Constants

Electric	Gravitational
q (charge)	m (mass)
ρ (volume charge density)	ρ (volume mass density)
σ (surface charge density)	σ (surface mass density)
λ (line charge density)	λ (line mass density)
$\mathbf{J}$ (convection current density)	$\mathbf{J}$ (mass current density)
$\mathbf{E}$ (electric field)	$\mathbf{g}$ (gravitational field)
$\mathbf{B}$ (magnetic field)	$\mathbf{K}$ (cogravitational field)
φ (scalar potential)	φ (scalar potential)
$\mathbf{A}$ (vector potential)	$\mathbf{A}$ (vector potential)
ε_0 (permittivity of space)	$-1/4\pi G$
μ_0 (permeability of space)	$-4\pi G/c^2$
$-1/4\pi\varepsilon_0$ or $-\mu_0 c^2/4\pi$	G (gravitational constant)
c (velocity of light)	c (velocity of gravitation)

It is important to keep in mind, however, that only electromagnetic equations for fields in a vacuum have their gravitational counterparts, and only the electromagnetic symbols listed in Table 11-1 can be directly replaced by the corresponding gravitational symbols. In all other cases the following conversion procedure should be used:

(1) If an electromagnetic equation is for fields in the presence of material media, reduce the equation to fields in a vacuum.

(2) If the equations contain field vectors $\mathbf{D}$ or $\mathbf{H}$, replace them by $\mathbf{E}$ or $\mathbf{B}$, using the relations $\mathbf{D} = \varepsilon_0\mathbf{E}$ and $\mathbf{B} = \mu_0\mathbf{H}$.

(3) Use Table 11-1 to replace electromagnetic symbols by the corresponding gravitational symbols.[4]

▼

Example 11-1.1 Using the analogy between electromagnetic and gravitational equations, convert Eqs. (3-2.6), (3-2.13), (4-1.13), (4-2.2), (4-3.34), (4-4.10), (4-5.2), (4-5.6), and (5-1.11).

Replacing in Eq. (3-2.6) **E** by **g** and **B** by **K**, we obtain for the cogravitational field associated with the gravitational field of a mass distribution moving with velocity **v**

$$\mathbf{K} = (\mathbf{v} \times \mathbf{g})/c^2. \tag{11-1.23}$$

Replacing in Eq. (3-2.13) **E** by **g** and **B** by **K**, we obtain for the cogravitational field associated with the gravitational field of a point mass in arbitrary motion

$$\mathbf{K} = \frac{\mathbf{r} \times \mathbf{g}}{cr}. \tag{11-1.24}$$

Replacing in Eq. (4-1.13) **E** by **g**, ε_0 by $-1/4\pi G$, and q by m, we obtain for the gravitational field of a uniformly moving point mass in terms of the present position of the mass[5, 6]

$$\mathbf{g} = -G\frac{m(1 - v^2/c^2)}{r_0^3[1 - (v^2/c^2)\sin^2\theta]^{3/2}}\mathbf{r}_0. \tag{11-1.25}$$

Replacing in Eq. (4-2.2) **H** by $\mathbf{B}/\mu_0$, **B** by **K**, μ_0 by $-4\pi G/c^2$, and q by m, we obtain for the cogravitational field of a uniformly moving point mass in terms of the present position of the mass

$$\mathbf{K} = -G\frac{m(1 - v^2/c^2)}{c^2 r_0^3\{1 - (v^2/c^2)\sin^2\theta\}^{3/2}}[\mathbf{v} \times \mathbf{r}_0]. \tag{11-1.26}$$

Replacing in Eq. (4-3.34) **E** by **g**, ε_0 by $-1/4\pi G$, and q by m, we obtain for the gravitational field of a point mass moving with acceleration (in terms of the retarded position of the mass)

$$\mathbf{g} = -G\frac{m}{r^3(1 - \mathbf{r}\cdot\mathbf{v}/rc)^3}\left\{\left(\mathbf{r} - \frac{r\mathbf{v}}{c}\right)\left(1 - \frac{v^2}{c^2}\right) + \mathbf{r} \times \left[\left(\mathbf{r} - \frac{r\mathbf{v}}{c}\right) \times \frac{\dot{\mathbf{v}}}{c^2}\right]\right\}. \tag{11-1.27}$$

Replacing in Eq. (4-4.10) **H** by $\mathbf{B}/\mu_0$, **B** by **K**, μ_0 by $-4\pi G/c^2$, and q by m, we obtain for the cogravitational field of a point mass moving with acceleration (in terms of the retarded position of the mass)

$$\mathbf{K} = -G\frac{m}{cr^2(1-\mathbf{r}\cdot\mathbf{v}/rc)^2}\left[\frac{1-v^2/c^2+\mathbf{r}\cdot\dot{\mathbf{v}}/c^2}{r(1-\mathbf{r}\cdot\mathbf{v}/rc)}(\mathbf{v}\times\mathbf{r})+\frac{\dot{\mathbf{v}}\times\mathbf{r}}{c}\right]. \tag{11-1.28}$$

Replacing in Eq. (4-5.2) ε_0 by $-1/4\pi G$ and q by m, we obtain for the gravitational scalar potential of a uniformly moving point mass (in terms of the present position of the mass)

$$\varphi = -G\frac{m}{r_0[1-(v^2/c^2)\sin^2\theta]^{1/2}}. \tag{11-1.29}$$

Replacing in Eq. (4-5.6) μ_0 by $-4\pi G/c^2$ and q by m, we obtain for the cogravitational vector potential of a uniformly moving point mass (in terms of the present position of the mass)

$$\mathbf{A} = -G\frac{m\mathbf{v}}{c^2r_0[1-(v^2/c^2)\sin^2\theta]^{1/2}}. \tag{11-1.30}$$

Replacing in Eq. (5-1.11) **E** by **g** and ε_0 by $-1/4\pi G$, we obtain for the gravitational field of a mass distribution of density ρ moving with constant speed (in terms of the present position of the mass)

$$\mathbf{g} = G\int\frac{\nabla'\rho-\mathbf{i}(v^2/c^2)(\partial\rho)/(\partial x')}{\{x_0'^2+(y'^2+z'^2)/\gamma^2\}^{1/2}}dV'. \tag{11-1.31}$$

Example 11-1.2 Consider a planet moving in a circular orbit about a central body. Using the electric field obtained in Example 4-4.1, discuss the consequences of retardation on the motion of the central body and on the planets whose orbits are interior relative to the orbit of the planet under consideration.[7]

Replacing in Eq. (4-4.39) **E** by **g**, ε_0 by $-1/4\pi G$, and q by m, we obtain for the gravitational field produced by the planet at the center of the orbit

$$\mathbf{g} = -G\frac{m}{r^3}\left\{\left(1 - \frac{v^2}{2c^2}\right)\mathbf{r}_0 - \frac{2rv^2}{3c^3}\mathbf{v}_0\right\}. \qquad (11\text{-}1.32)$$

According to Eq. (11-1.32), the gravitational field produced by the planet is quite different from Newton's gravitational field. In particular, because of the presence of the component in the direction of the instantaneous velocity vector, the field is not even radial. In our solar system, this new component of the gravitational field may have important consequences both on the motion of the Sun and on the motion of planets. Although the field given by Eq. (11-1.32) is for the center of the orbit, this field should be approximately correct within a certain region of space around the center of the orbit. As far as the Sun is concerned, the new component of the gravitational field exerts then a torque on the Sun and causes it to rotate in the direction of the orbital velocity of the planet.

Because of the new component of the gravitational field, outer planets should produce a similar effect on the motion of the inner planets, causing an acceleration (and deceleration) of their orbital velocities and, what is most important, causing a secular motion of the large axes of the orbits of the inner planets in the direction of the orbital velocity of the outer planets.

In the middle of the last century, Urbain Le Verrier found that Newton's gravitational law was incapable of explaining certain discrepancies between the observed and calculated parameters of planetary motions. In particular, he computed the secular perturbations of the motion of Mercury under the action of the other planets and found that there was an inexplicable "residual" precession of Mercury's perihelion. According to the presently accepted data, the precession of Mercury's perihelion is approximately 575 seconds of arc per century, of which 532 seconds can be attributed to Newtonian attraction between Mercury and other planets, while about 43 seconds cannot be explained on the basis of Newton's gravitational law. It was the greatest triumph of Einstein's general relativity theory when, on the basis of this theory, Einstein explained the residual 43 seconds in the precession of Mercury's perihelion. In fact, to this day most of the credibility of the general relativity theory is directly attributable to the amazing accuracy of this explanation and therefore indirectly attributable to the accuracy of celestial mechanics based on Newton's gravitational law.

But according to Eq. (11-1.32), the precession of Mercury's perihelion caused by other planets may be different from the presently accepted 532 seconds. Furthermore, the gravitational field experienced by a planet in the reference frame of the planet is not a Newtonian field, but the field given by Eq. (11-1.25). And in the heliocentric reference frame there is also a cogravitational field created by the rotating Sun. Therefore the true "residual" precession (if it exists at all) may be quite different from the presently accepted 43 seconds. Thus the explanation of the residual precession of Mercury by the general relativity theory can hardly be considered as definitive.

▲

11-2. Relativistic Transformation Equations for Gravitational and Cogravitational Fields

In Chapters 6 and 7 we derived relativistic transformation equations for electric and magnetic fields starting from Eq. (5-1.11) representing the electric field of a uniformly moving charge distribution and Eq. (3-2.6) expressing the magnetic field of a uniformly moving charge distribution in terms of its electric field. As was shown in Example 11-1.1, the gravitational counterparts of these equations are Eqs. (11-1.31) and (11-1.23), which differ from the electromagnetic equations only by the field symbols and by the factor and sign in front of the integral in Eq. (11-1.31) and in the corresponding time-independent equation. Therefore the same calculations that led to the relativistic transformation equations for electromagnetic fields can be duplicated for deriving relativistic transformation equations for the gravitational-cogravitational fields. The only difference between the resulting equations and the corresponding electromagnetic equations will then be the appearance of the components of **g** and **K** in the gravitational-cogravitational equations instead of the components of **E** and **B** in the corresponding electromagnetic equations.[8]

The same holds for the relativistic transformation equations for potentials (see Section 6-4), except that in the case of potentials there is no need to change the symbols.

Thus there is actually no need to *derive* the relativistic transformation equations for the gravitational-cogravitational fields and potentials. All we need to do for obtaining these equations is to replace the components **E** and **B** in Eqs. (7-1.5)-(7-1.14) and (7-1.23)-(7-1.32) by the corresponding components of **g** and **K** and to copy Eqs. (7-1.1)-(7-1.4), (7-1.19)-(7-1.22), (7-1.15)-(7-1.18) and (7-1.33)-(7-1.36). The resulting relativistic transformation equations for the quantities measured in Σ expressed in terms of the quantities measured in Σ′ are:

(a) For the space and time coordinates (these equations are the same as those derived in Chapters 6 and 7 by considering electric and magnetic fields)

$$x = \gamma(x' + vt'), \tag{11-2.1}$$

$$y = y', \tag{11-2.2}$$

$$z = z', \tag{11-2.3}$$

$$t = \gamma(t' + vx'/c^2). \tag{11-2.4}$$

(b) For the gravitational field

$$g_x = g_x', \tag{11-2.5}$$

$$g_y = \gamma(g_y' + vK_z'), \tag{11-2.6}$$

$$g_z = \gamma(g_z' - vK_y'). \tag{11-2.7}$$

(c) For the cogravitational field

$$K_x = K_x', \tag{11-2.8}$$

$$K_y = \gamma(K_y' - vg_z'/c^2), \tag{11-2.9}$$

$$K_z = \gamma(K_z' + vg_y'/c^2). \tag{11-2.10}$$

(d) For the mass and mass current densities

$$\rho = \gamma[\rho' + (v/c^2)J_x'], \tag{11-2.11}$$

$$J_x = \gamma(J_x' + v\rho'), \tag{11-2.12}$$

$$J_y = J_y', \tag{11-2.13}$$

$$J_z = J_z'. \tag{11-2.14}$$

(e) For the gravitational scalar potential and the cogravitational vector potential

$$\varphi = \gamma(\varphi' + vA_x'), \tag{11-2.15}$$

$$A_x = \gamma[A_x' + (v/c^2)\varphi'], \tag{11-2.16}$$

$$A_y = A_y', \tag{11-2.17}$$

$$A_z = A_z'. \tag{11-2.18}$$

Relativistic equations for the quantities measured in Σ' expressed in terms of the quantities measured in Σ are:

(a) For the space and time coordinates (these equations are the same as those derived in Chapters 6 and 7 by considering electric and magnetic fields)

$$x' = \gamma(x - vt), \tag{11-2.19}$$

$$y' = y, \tag{11-2.20}$$

$$z' = z, \tag{11-2.21}$$

$$t' = \gamma(t - vx/c^2). \tag{11-2.22}$$

(b) For the gravitational field

$$g'_x = g_x, \tag{11-2.23}$$

$$g'_y = \gamma(g_y - vK_z), \tag{11-2.24}$$

$$g'_z = \gamma(g_z + vK_y). \tag{11-2.25}$$

(c) For the cogravitational field

$$K'_x = K_x, \tag{11-2.26}$$

$$K'_y = \gamma(K_y + vg_z/c^2), \tag{11-2.27}$$

$$K'_z = \gamma(K_z - vg_y/c^2). \tag{11-2.28}$$

(d) For the mass and mass current densities

$$\rho' = \gamma[\rho - (v/c^2)J_x], \tag{11-2.29}$$

$$J'_x = \gamma(J_x - v\rho), \tag{11-2.30}$$

$$J'_y = J_y, \tag{11-2.31}$$

$$J'_z = J_z. \tag{11-2.32}$$

(e) For the gravitational scalar potential and the cogravitational vector potential

$$\varphi' = \gamma(\varphi - vA_x), \tag{11-2.33}$$

$$A'_x = \gamma[A_x - (v/c^2)\varphi], \tag{11-2.34}$$

$$A_y' = A_y, \qquad (11\text{-}2.35)$$

$$A_z' = A_z. \qquad (11\text{-}2.36)$$

Quite clearly, transformation equations obtained in Chapter 7 for quantities not containing electric and magnetic fields or their components (such as velocity, acceleration, gradient, etc.) remain valid for gravitational-cogravitational fields as well.

▼

Example 11-2.1 The Newtonian equation for the gravitational field of a stationary point mass is

$$\mathbf{g} = -G\frac{m}{r^3}\mathbf{r}. \qquad (11\text{-}2.37)$$

Starting with this equation and using relativistic transformation obtain the equation for the gravitational field of a point mass moving with uniform velocity v parallel to the x axis.

For simplicity, let us assume that the gravitational field is in the xy plane. In this case r in Eq. (11-2.37) is $r = (x^2 + y^2)^{1/2}$.

To obtain the gravitational field of the mass when the mass moves with constant speed parallel to the x axis, we shall assume that the mass is located in a reference frame Σ' which moves with velocity $\mathbf{v} = v\mathbf{i}$ relative to the laboratory (reference frame Σ). By Eq. (11-2.37), in the reference frame Σ' the x component of the field is given by

$$g_x' = -G\frac{m}{(x'^2 + y'^2)^{3/2}}x', \qquad (11\text{-}2.38)$$

and the y component is given by

$$g_y' = -G\frac{m}{(x'^2 + y'^2)^{3/2}}y'. \qquad (11\text{-}2.39)$$

Since the mass is stationary in Σ', we are free to choose the time of observation in Σ. We choose $t = 0$. Equation (11-2.5) tells us that to find g_x of the moving mass, we must replace g_x' on the left of Eq. (11-2.38) by g_x, while Eq. (11-2.19) tells us that, since $t = 0$, we must replace x' in Eq. (11-2.38) by γx [observe that in Eq. (11-2.38) x appears in the numerator and in the denominator]. Finally, Eq. (11-2.20) tells us that y' in the denominator of Eq. (11-2.19) must be replaced by y. Making the substitutions, we obtain for g_x of the moving point mass

$$g_x = -G\frac{m}{[(\gamma x)^2 + y^2]^{3/2}}\gamma x = -G\frac{m}{\gamma^2(x^2 + y^2/\gamma^2)^{3/2}}x. \quad (11\text{-}2.40)$$

To obtain the y component of the field of the moving mass, we shall use Eqs. (11-2.6), (11-2.19), and again Eq. (11-2.20). Since $\mathbf{K} = 0$ for the stationary mass, Eq. (11-2.6) tells us that, to find g_y of the moving mass, we must replace g_y' on the left of Eq. (11-2.39) by g_y/γ, while Eqs. (11-2.19) and (11-2.20) tell us that we must replace x' in Eq. (11-2.39) by γx and y' by y. Making the substitutions, we then obtain for g_y of the moving point mass

$$g_y/\gamma = -G\frac{m}{[(\gamma x)^2 + y^2]^{3/2}}y, \quad (11\text{-}2.41)$$

or

$$g_y = -G\frac{m}{\gamma^2(x^2 + y^2/\gamma^2)^{3/2}}y. \quad (11\text{-}2.42)$$

Replacing now γ in Eqs. (11-2.40) and (11-2.42) by $1/(1 - v^2/c^2)^{1/2}$, factoring out $x^2 + y^2$ from the denominator, taking into account that $\mathbf{r} = x\mathbf{i} + y\mathbf{j}$, where $\mathbf{i}$ and $\mathbf{j}$ are unit vectors in the direction of the x and y axes, and noting that $y^2/(x^2 + y^2) = \sin^2\theta$, where θ is the angle between $\mathbf{v}$ and $\mathbf{r}$, we finally obtain

$$\mathbf{g} = -G\frac{m(1 - v^2/c^2)}{r^3[1 - (v^2/c^2)\sin^2\theta]^{3/2}}\mathbf{r}, \quad (11\text{-}2.43)$$

which is the same equation (the "Heaviside equation") that we obtained in Example 11-1.1 [Eq. (11-1.25)] by transforming electromagnetic equations into gravitational equations [r and $\mathbf{r}$ in

Eq. (11-2.43) represent the present position of the mass and are therefore the same as r_0 and $\mathbf{r}_0$ in Eq. (11-1.25)].

Note that in applying relativistic transformations we did not transform the mass m. Just like the electric charge q, the gravitational mass of a body is invariant under relativistic transformations. In fact, the inertial mass is also invariant, as was explained in Section 8-4.

▲

11-3. Relativistic Gravitation and Relativistic Mechanics

In Chapter 8 we developed relativistic mechanics on the basis of the force, momentum, and energy relations pertaining to electromagnetic fields. Now we need to determine whether the same relativistic mechanics applies to gravitational interaction as well.

First we note that the gravitational counterpart of the Lorentz force law is[9]

$$\mathbf{F} = m(\mathbf{g} + \mathbf{u} \times \mathbf{K}), \tag{11-3.1}$$

where $\mathbf{F}$ is the force acting on a point mass m moving with velocity $\mathbf{u}$ in the presence of a gravitational field $\mathbf{g}$ and a cogravitational field $\mathbf{K}$. This force law is invariant under relativistic transformations. Examining the calculations used in Section 8-1 for obtaining relativistic transformations of the force on the basis of Lorentz force law, we recognize that exactly the same calculations can be used for obtaining the same force transformation equations on the basis of Eq. (11-3.1), except that instead of the components of $\mathbf{E}$ and $\mathbf{B}$ we must then use the components of $\mathbf{g}$ and $\mathbf{K}$. Thus the relativistic force transformation equations obtained in Chapter 8 are valid for the gravitational interactions as was to be expected, since gravitational fields, just like the electromagnetic fields, are force fields.

Consider now the transformation equations for energy and momentum obtained in Section 8-2. All three electromagnetic equations, Eqs. (8-2.1), (8-2.2), and (8-2.3), that we used for obtaining the transformation equations for energy and momentum have their gravitational-cogravitational counterparts that can be obtained from the electromagnetic equations by means of substitutions listed in Table 11-1.[10] Therefore the equations derived in Section 8-2 can be converted into equations for gravitational-cogravitational fields by substitution of symbols in accordance with Table 11-1.

In particular, from Eq. (8-2.11) we have for the energy of interaction U_g' between a point mass m and the gravitational field (the gravitational potential energy of m) measured in Σ' and expressed in terms of the gravitational interaction energy U_g and the cogravitational interaction energy U_c' of m measured in Σ

$$U_g' = \gamma(U_g - U_c), \tag{11-3.2}$$

and, from Eq. (8-2.12), the inverse relation

$$U_g = \gamma(U_g' + U_c'). \tag{11-3.3}$$

Just as in the case of the electromagnetic interaction, Eqs. (11-3.2) and (11-3.3) can be expressed in terms of the gravitational-cogravitational field momentum **G** as [see Eqs. (8-2.13) and (8-2.14)]

$$U_g' = \gamma(U_g - vG_x) \tag{11-3.4}$$

and

$$U_g = \gamma(U_g' + vG_x'). \tag{11-3.5}$$

Similarly, from Eqs. (8-2.18)-(8-2.23) we obtain the transformation equations for the gravitational-cogravitational field momentum

$$G_x' = \gamma[G_x - (v/c^2)U_g], \tag{12-3.6}$$

$$G_y' = G_y, \tag{11-3.7}$$

$$G_z' = G_z; \tag{11-3.8}$$

and

$$G_x = \gamma[G_x' + (v/c^2)U_g'], \tag{11-3.9}$$

$$G_y = G_y', \tag{11-3.10}$$

$$G_z = G_z'. \tag{11-3.11}$$

Except for the subscripts in Eqs. (11-3.2)-(11-3.9), the above transformation equations for gravitational-cogravitational energy and momentum are the same as those for the electromagnetic energy and momentum. Therefore all the consequences deduced from the electromagnetic energy and momentum transformation equations in Chapter 8, Sections 3-7, remain unchanged.

Thus, provided that the velocity of propagation of gravitational and cogravitational fields is the same as the velocity of light, our relativistic transformation equations for gravitational and cogravitational fields confirm the general validity of the transformation equations for mechanical force, energy, and momentum derived in Chapter 8. If the velocity of propagation of gravitational and cogravitational fields is not the same as the velocity of light, our relativistic transformation equations would still remain correct but, because the force and momentum equations would then contain the constant c different from c appearing in the corresponding electromagnetic equations, the mechanical behavior of rapidly moving bodies responding to gravitational interactions would be different from the behavior of rapidly moving bodies responding to electromagnetic interactions. In effect, there would be two different mechanics – the "gravitational-cogravitational mechanics," and the "electromagnetic mechanics" – involving different effective masses, different effective momenta, and different rest energies.

A possibility exists that our relativistic transformation equations are not entirely correct. According to Einstein's mass-energy equation, any energy has a certain mass. But any mass is a source of gravitation. Therefore the gravitational field of a mass distribution may be caused not only by the mass of the distribution as such, but also by the gravitational energy of this distribution.[11] In this case the equation for the divergence of the gravitational field, Eq. (11-1.5), becomes only approximately correct, and all equations derived with the help of Eq. (11-1.5) become also only approximately correct. It is important to note, however, that this effect, if it exists, is extremely small.[12]

11-4. Covariant Formulation of the Electromagnetic and of the Gravitational-Cogravitational Theories[13]

In 1906, H. Poincaré discovered that Lorentz transformations of coordinates and time could be associated with an imaginary four-dimensional "space" represented by four "orthogonal" axes, three of which were the ordinary space axes while the fourth was the time axis calibrated in units of it, where $i = \sqrt{-1}$.[14] The effect of the Lorentz transformations applied to a position vector in this four-dimensional space could be interpreted as a change of the components of this vector caused by a rotation of the axes around the origin. The idea of associating relativistic transformations with geometrical relations in four-dimensional space was later developed by Hermann Minkowski,[15] who laid the foundation of what is known as the *covariant* formulation of electrodynamics.

Whereas in the standard formulation of electrodynamics the basic mathematical elements are scalars and vectors in ordinary three-dimensional space, in the covariant formulation the basic mathematical elements are scalars ("Lorentzian scalars"), *4-vectors*, and *4-tensors* in the four-dimensional space ("Minkowski space").

There are two commonly used notations for 4-vectors. In the so-called "covariant" notation a 4-vector A is identified by its four components written as A_1, A_2, A_3, A_4, or, for brevity, as A_μ with μ accepting the values of 1, 2, 3, and 4. In the so-called "contravariant" notation a 4-vector $\boldsymbol{A}$ is identified by its four components written as A^0, A^1, A^2, A^3, or, for brevity, as A^μ with μ accepting the values of 0, 1, 2, and 3.[16] We shall only use the covariant notation and shall designate 4-vectors by bold italic letters. In the covariant notation the first three components of a 4-vector are the components along the ordinary space axes while the fourth component is along the time axis.

The exact definition of what particular entity constitutes a 4-vector starts with the definition of the "position 4-vector" $\boldsymbol{r}$ in the laboratory reference frame Σ as

$$\boldsymbol{r} = (x_1, x_2, x_3, x_4) = (x, y, z, ict) = (\mathbf{r}, ict), \qquad (11\text{-}4.1)$$

where $\mathbf{r}$ inside the last parentheses is used as an abbreviation for the three *components* of $\boldsymbol{r}$ along the actual space axes.[17] In the moving reference frame Σ' the position vector is then

$$\boldsymbol{r}' = (x_1', x_2', x_3', x_4') = (x', y', z', ict') = (\mathbf{r}', ict'). \qquad (11\text{-}4.2)$$

If one applies Lorentz transformations of coordinates and time to Eq. (11-4.2), one obtains (see Example 11-4.1)

$$\boldsymbol{r}' = [\gamma(x_1 + i\beta x_4), x_2, x_3, \gamma(x_4 - i\beta x_1)], \qquad (11\text{-}4.3)$$

where $\beta = v/c$. But since the effect of Lorentz transformations on the components of $\boldsymbol{r}$ is the same as a rotation of the axes in the four-dimensional space, and since the rotation of the axes should affect the components of all 4-vectors in a similar way, one defines any 4-vector

$$\boldsymbol{A} = (A_1, A_2, A_3, A_4) = (\mathbf{A}, ic\alpha) \qquad (11\text{-}4.4)$$

as an entity that transforms under a relativistic transformation ("Lorentz transformation"), just like $\boldsymbol{r}$ does, into

$$\boldsymbol{A}' = [\gamma(A_1 + i\beta A_4), A_2, A_3, \gamma(A_4 - i\beta A_1)]. \qquad (11\text{-}4.5)$$

Now, Eq. (11-4.5) can be written as

$$\boldsymbol{A}' = (A_1', A_2', A_3', A_4') = (\mathbf{A}', ic\alpha'). \tag{11-4.6}$$

Thus a 4-vector, by its very definition, retains its form under the Lorentz transformation, or, as one says, is "Lorentz covariant," or "space-time covariant," or simply "covariant." Therefore any physical law expressed as a relation between 4-vectors remains the same in all uniformly moving reference frames and thus automatically satisfies the principle of relativity.

An example of an electromagnetic 4-vector is the electromagnetic "current density" 4-vector

$$\boldsymbol{J} = (J_1, J_2, J_3, J_4) = (J_x, J_y, J_z, ic\rho). \tag{11-4.7}$$

Observe that $\boldsymbol{J}$ incorporates the components of the ordinary current density $\mathbf{J}$ and the charge density ρ into a single entity.

Not all frame-independent physical quantities can be incorporated into 4-vectors. In particular, it is impossible to express the electric and magnetic field vectors in the form of 4-vectors. However, electric and magnetic fields can be incorporated into the covariant "electromagnetic 4-tensor" compatible with the 4-vector formulation of the current density and of other electromagnetic quantities. This electromagnetic 4-tensor is designated as $F_{\mu\nu}$ and is defined as

$$\boldsymbol{F}_{\mu\nu} = \begin{bmatrix} 0 & B_z & -B_y & -iE_x/c \\ -B_z & 0 & B_x & -iE_y/c \\ B_y & -B_x & 0 & -iE_z/c \\ iE_x/c & iE_y/c & iE_z/c & 0 \end{bmatrix}, \tag{11-4.8}$$

where the subscript μ indicates the row (1, 2, 3, 4 top to bottom) and the subscript ν indicates the column (1, 2, 3, 4 left to right).

As an example of the use and applications of the covariant formulation of electromagnetism, consider the equation

$$\sum_{\nu=1}^{4} \frac{\partial F_{\mu\nu}}{\partial x_\nu} = \mu_0 J_\mu . \tag{11-4.9}$$

With $\mu = 4$ (see Example 11-4.2) this equation represents Maxwell's equation (in terms of its Cartesian components)

$$\nabla \cdot \mathbf{E} = \frac{\rho}{\varepsilon_0} . \tag{11-4.10}$$

With $\mu = 1, 2, 3$ (see Example 11-4.2) the same equation represents Maxwell's equation (in terms of its Cartesian components)

$$\nabla \times \mathbf{B} - \frac{1}{c^2}\frac{\partial \mathbf{E}}{\partial t} = \mu_0 \mathbf{J} . \tag{11-4.11}$$

Likewise, the equation (see Example 11-4.3)

$$\frac{\partial F_{\mu\nu}}{\partial x_\lambda} + \frac{\partial F_{\nu\lambda}}{\partial x_\mu} + \frac{\partial F_{\lambda\mu}}{\partial x_\nu} = 0 , \tag{11-4.12}$$

where $\mu \neq \nu \neq \lambda = 1, 2, 3, 4$ represents the remaining two Maxwell's equations

$$\nabla \cdot \mathbf{B} = 0 \tag{11-4.13}$$

and

$$\nabla \times \mathbf{E} = -\frac{\partial \mathbf{B}}{\partial t} . \tag{11-4.14}$$

Covariant formulation is considered by some authors to be the most appropriate formulation for expressing the laws of physics in a frame-independent form. It is also believed to be by some authors more concise and occasionally more informative than the conventional formulation. Since any equation and relation consistent with the principle of relativity should be expressible in a covariant form, and since the principle of relativity is considered to be a fundamental law of nature, the laws of physics that cannot be expressed in a covariant form are considered by some authors to be incomplete or incorrect.[18]

Newton's gravitational law is an example of a physical law that cannot be expressed in a covariant form. The problem of finding an invariant form of the law of gravitation was first considered by Poincaré, but without success.[19] It is interesting to note that Poincaré attempted to solve the problem on the basis of just one gravitational field (the gravitational analog of the electrostatic field). But even if the theory of gravitation is built upon two fields, as we have done in this chapter, a covariant theory of gravitation is not possible unless the gravitational mass, just like the electric charge, does not depend on the velocity with which the mass moves.

Until recently it was generally believed that the mass of a moving body was a function of the velocity of the body (see Section 8-4) and thus was not invariant under relativistic transformations. The alleged noninvariance of mass under relativistic transformations was therefore the most important reason for rejecting a theory of gravitation analogous to the theory of electromagnetism: if mass, unlike the electric charge, is not invariant, then the analogy between electromagnetism and gravitation cannot be complete, a covariant formulation of gravitation is impossible, and therefore no gravitational theory based on the gravitational field vector, with or without the addition of a second (the cogravitational) field vector, can be correct. A theory of gravitation was therefore created by Einstein based not on the concept of the gravitational force field, but on the concept of the "curvature of space."[20]

However, as we now know, neither the gravitational nor the inertial mass depends on the velocity with which a body moves. In fact, as far as the principle of relativity, relativistic transformations of all pertinent quantities, and relativistic mechanics are concerned, the analogy between electromagnetism and gravitation-cogravitation is complete. Therefore a covariant formulation of the theory of gravitation based on gravitational-cogravitational fields is not only possible but can be constructed straightaway from the covariant theory of electromagnetism by a mere substitution of symbols and constants in accordance with Table 11-1.

Thus, for example, the 4-vector mass current can be obtained from Eq. (11-4.7) [according to Table 11-1, none of the symbols or constants in Eq. (11-4.7) need to be replaced]; the result is

$$\boldsymbol{J} = (J_1, J_2, J_3, J_4) = (J_x, J_y, J_z, ic\rho). \qquad (11\text{-}4.15)$$

Likewise, the gravitational-cogravitational field tensor can be obtained from Eq. (11-4.8), this time by replacing, with the help of Table 11-1, the components of **E** by the corresponding components of **g** and the components of **B** by the corresponding components of **K**; the result is

$$\boldsymbol{F}_{\mu\nu} = \begin{bmatrix} 0 & K_z & -K_y & -ig_x/c \\ -K_z & 0 & K_x & -ig_y/c \\ K_y & -K_x & 0 & -ig_z/c \\ ig_x/c & ig_y/c & ig_z/c & 0 \end{bmatrix}. \qquad (11\text{-}4.16)$$

Finally, from Eqs. (11-4.9), (11-4.12) and Table 11-1 we obtain for the basic laws of gravitational-cogravitational fields

$$\sum_{\nu=1}^{4} \frac{\partial F_{\mu\nu}}{\partial x_\nu} = -\frac{4\pi G}{c^2} J_\mu \qquad (11\text{-}4.17)$$

and

$$\frac{\partial F_{\mu\nu}}{\partial x_\lambda} + \frac{\partial F_{\nu\lambda}}{\partial x_\mu} + \frac{\partial F_{\lambda\mu}}{\partial x_\nu} = 0. \qquad (11\text{-}4.18)$$

It should be kept in mind, however, that c in the gravitational equations stands for the speed of propagation of gravitational-cogravitational fields, which is generally assumed to be the same as the speed of light, but has never been actually measured.

▼

Example 11-4.1 Starting with Eq. (11-4.2) derive Eq. (11-4.3).

Applying Eqs. (7-1.19)-(7-1.22) to the x', y', z', and ict' components of r' in Eq. (11-4.2), we obtain

$$\begin{aligned} r' &= [\gamma(x - vt), y, z, ic\gamma(t - vx/c^2)] \\ &= [\gamma(x - vict/ic), y, z, \gamma(ict - icvx/c^2)] \\ &= \{\gamma[x + i(v/c)ict], y, z, \gamma[ict - i(v/c)x]\}. \end{aligned} \tag{11-4.19}$$

Replacing in Eq. (11-4.19) x, y, z, and ict by x_1, x_2, x_3, x_4, and replacing v/c by β, we obtain Eq. (11-4.3).

Example 11-4.2 Show that Eq. (11-4.9) is equivalent to Eqs. (11-4.10) and (11-4.11).

Replacing in Eq. (11-4.9) $F_{\mu\nu}$ by $F_{4\nu}$, substituting x, y, z, and ict for x_1, x_2, x_3, and x_4, respectively, using, according to Eq. (11-4.8), $F_{41} = iE_x/c$, $F_{42} = iE_y/c$, $F_{43} = iE_z/c$, and $F_{44} = 0$, and using, according to Eq. (11-4.7), $J_4 = ic\rho$, we have

$$\frac{\partial(iE_x/c)}{\partial x} + \frac{\partial(iE_y/c)}{\partial y} + \frac{\partial(iE_z/c)}{\partial z} + \frac{\partial 0}{\partial(ict)} = \mu_0 ic\rho, \tag{11-4.20}$$

and since $\mu_0 c^2 = \varepsilon_0$, we obtain Eq. (11-4.10) (written in terms of Cartesian components).

Setting in Eq. (11-4.9) $\mu = 1$, we similarly obtain

$$\frac{\partial 0}{\partial x} + \frac{\partial B_z}{\partial y} - \frac{\partial B_y}{\partial z} - \frac{\partial(iE_x/c)}{\partial(ict)} = \mu_0 J_x, \tag{11-4.21}$$

or

$$\frac{\partial B_z}{\partial y} - \frac{\partial B_y}{\partial z} - \frac{\partial E_x}{c^2 \partial t} = \mu_0 J_x, \tag{11-4.22}$$

which is the x component of Eq. (11-4.11). Setting $\mu = 2$ and then $\mu = 3$, we obtain the y and z components of Eq. (11-4.11).

Example 11-4.3 Show that Eq. (11-4.12) is equivalent to Eqs. (11-4.13), and (11-4.14).

Setting in Eq. (11-4.12) $\mu = 1$, $\nu = 2$, $\lambda = 3$, and using Eq. (11-4.8), we obtain

$$\frac{\partial B_z}{\partial z} + \frac{\partial B_x}{\partial x} + \frac{\partial B_y}{\partial y} = 0, \tag{11-4.23}$$

which is the same as Eq. (11-4.13).

Setting in Eq. (11-4.12) $\mu = 2$, $\nu = 3$, $\lambda = 4$, and using Eqs. (11-4.8), we obtain

$$\frac{\partial B_x}{\partial (ict)} + \frac{\partial(-iE_z/c)}{\partial y} + \frac{\partial(iE_y/c)}{\partial z} = 0, \tag{11-4.24}$$

or

$$\frac{\partial E_z}{\partial y} - \frac{\partial E_y}{\partial z} = -\frac{\partial B_x}{\partial t}, \tag{11-4.25}$$

which is the x component of Eq. (11-4.14). The remaining two components are obtained in the same manner by setting $\mu = 1$, $\nu = 3$, $\lambda = 4$ and $\mu = 1$, $\nu = 2$, $\lambda = 4$.

Example 11-4.4 Show that Eq. (11-4.17) is equivalent to Eqs. (11-1.5), and (11-1.8) and that Eq. (11-4.18) is equivalent to Eqs. (11-1.6), and (11-1.7).

Replacing in Eq. (11-4.17) $F_{\mu\nu}$ by $F_{4\nu}$, substituting x, y, z, and ict for x_1, x_2, x_3, and x_4, respectively, using, according to Eq. (11-4.16), $F_{41} = ig_x/c$, $F_{42} = ig_y/c$, $F_{43} = ig_z/c$, and $F_{44} = 0$, and using, according to Eq. (11-4.15), $J_4 = ic\rho$, we have

$$\frac{\partial(ig_x/c)}{\partial x} + \frac{\partial(ig_y/c)}{\partial y} + \frac{\partial(ig_z/c)}{\partial z} + \frac{\partial 0}{\partial(ict)} = -\frac{4\pi G}{c^2} ic\rho, \tag{11-4.26}$$

which, after cancelling i and c, becomes the same as Eq. (11-1.5).

Setting in Eq. (11-4.17) $\mu = 1$, and using, according to Eq. 11-4.16), $F_{11} = 0$, $F_{12} = K_z$, $F_{13} = -K_y$, $F_{14} = -ig_x/c$, we similarly obtain

$$\frac{\partial 0}{\partial x} + \frac{\partial K_z}{\partial y} - \frac{\partial K_y}{\partial z} - \frac{\partial (ig_x/c)}{\partial (ict)} = \frac{4\pi G}{c^2}J_x, \qquad (11\text{-}4.27)$$

or

$$\frac{\partial K_z}{\partial y} - \frac{\partial K_y}{\partial z} = -\frac{4\pi G}{c^2}J_x + \frac{\partial g_x}{c^2 \partial t}, \qquad (11\text{-}4.28)$$

which is the x component of Eq. (11-1.8). Likewise, setting $\mu = 2$ and then $\mu = 3$ in Eq. (11-4.17) and using Eq. (11-4.16), we obtain the y and z components of Eq. (11-4.11).

Setting in Eq. (11-4.18) $\mu = 1$, $\nu = 2$, $\lambda = 3$, and using Eq. (11-4.16), we obtain

$$\frac{\partial K_z}{\partial z} + \frac{\partial K_x}{\partial x} + \frac{\partial K_y}{\partial y} = 0, \qquad (11\text{-}4.29)$$

which is the same as Eq. (11-1.6).

Setting in Eq. (11-4.18) $\mu = 2$, $\nu = 3$, $\lambda = 4$, and using Eq. (11-4.16), we similarly obtain

$$\frac{\partial g_z}{\partial y} - \frac{\partial g_y}{\partial z} = -\frac{\partial K_x}{\partial t}, \qquad (11\text{-}4.30)$$

which is the x component of Eq. (11-1.7). The remaining two components are obtained in the same manner by setting $\mu = 1$, $\nu = 3$, $\lambda = 4$ and $\mu = 1$, $\nu = 2$, $\lambda = 4$.

▲

References and Remarks for Chapter 11

1. See Oleg D. Jefimenko, *Causality, Electromagnetic Induction, and Gravitation* (Electret Scientific, Star City, 1992), Chapters 4-8.

2. The existence of a second gravitational field, similar to the magnetic field, was first suggested by Oliver Heaviside in his two-part article "A Gravitational and Electromagnetic Analogy," The Electrician **31**, 281-282 and 359 (1893).

3. Although most of these equations are new, the idea of developing Newton's gravitational theory in a manner analogous to the electromagnetic theory is not new (see Ref. 2). Unfortunately, it was abandoned at its very genesis because of the rapid and forceful development of Einstein's relativity theories, which have imposed severe restrictions on what is considered by many to constitute "competent" scientific work. In fact, Einstein's general relativity theory is considered by some scientists to be the definitive theory of gravitation, making all alternative gravitational theories either superfluous or "unscientific." However, the recently discovered retarded equations for gravitational and cogravitational fields [see Ref. 1 and Eqs. (11-1.9) and (11-1.10) (causal gravitational equations); see also Example 11-1.2] point out a path for an unquestionably legitimate new inquiry into the nature and properties of gravitational fields and interactions. Until this path is fully explored, one cannot accept any gravitational theory as "definitive."

4. Many gravitational equations obtained by transforming the corresponding electromagnetic equations are given in Ref. 1, pp. 106-111.

5. This equation was first derived by Heaviside in 1893. See Ref.2.

6. For a direct derivation of this equation see Oleg D. Jefimenko, "Gravitational Field of a Point Mass Moving with Uniform Linear or Circular Velocity," Galilean Electrodynamics **5**, 25-33 (1994).

7. This problem is discussed in detail in Ref. 6.

8. See also Oleg D. Jefimenko, "Derivation of Relativistic Transformations for Gravitational Fields from Retarded Field Integrals," Galilean Electrodynamics **6**, 23-30 (1995).

9. See Ref. 1, pp. 80-84 and 128-129.

10. See also Ref. 1, pp. 106-111.

11. For a detailed discussion of this effect, including the possibility of antigravitational mass distributions arising from it, see Ref. 1 pp, 140-158.

12. It is possible that relativistic electrodynamics is also only approximately correct, since it is valid only for electromagnetic fields in a perfect vacuum. However, there is some evidence that a perfect vacuum does not exist, and that electromagnetic fields in "empty" space are affected by what is known as "polarization of the vacuum."

13. The purpose of this section is merely to make clear that the relativistic theory of gravitation outlined in the preceding sections of this chapter can be expressed in a covariant form without any modifications or additions. Only the most basic information on the covariant formulation is presented in this section. For detailed expositions of covariant electrodynamics the reader is referred to such books as E. J. Konopinski, *Electromagnetic Fields and Relativistic Particles*, (McGraw-Hill, New York, 1981); J. D. Jackson, *Classical Electrodynamics*, 2nd. ed., (Wiley, New York, 1975); F. Sauter and R. Becker, *Electromagnetic Fields and Interactions*, (Blaisdell, New York, 1964); W. K. H. Panofsky and M. Phillips, *Classical Electricity and Magnetism*, 2nd ed., (Addison-Wesley, Reading, Massachusetts, 1962). For an elementary introduction to 4-vectors see W. G. V. Rosser, *Introductory Relativity*, (Plenum, New York, 1967).

14. H. Poincaré, "Sur la Dynamique de L'Électron," Rend. Circ. mat. Palermo **21**, 129-176 (1906).

15. H. Minkowski, "Die Grundlagen für die elektromagnetischen Vorgänge in bewegter Körpern," Göttinger Nachr. 53-111 (1908).

16. In the contravariant representation of 4-vectors the time axis is calibrated in units of *t* rather than in units of *it*, the sequence of the components of the 4-vectors is denoted as 0, 1, 2, and 3, with the *t* component being the 0 component.

17. They may or may not be equal to the Cartesian components of the corresponding vectors in the three-dimensional space.

18. Some authors consider this view as extreme, since, according to it, even Maxwell's equations in their vector form should be classified as "incomplete" or " incorrect" (see Section 7.4). Note also that covariant formulation of equations changes the form of the equations but does not create new physical laws.

19. See Ref. 12 and H. Poincaré, "La Dynamique de L'Électron," in *Revue général des Sciences pures et appliquées* **19**, 386-402 (1908).

20. A. Einstein, "Grundlage der allgemeinen Relativitätstheorie," Ann. Pys. **49**, 769-822 (1916).

APPENDIX

VECTOR IDENTITIES

In the vector identities listed below φ and U are scalar point functions; **A**, **B**, and **C** are vector point functions; X is a scalar or vector point function of primed coordinates and incorporates an appropriate multiplication sign (dot or cross for vectors).

Box product

(V-1) $$\mathbf{A}\cdot(\mathbf{B}\times\mathbf{C}) = \mathbf{B}\cdot(\mathbf{C}\times\mathbf{A}) = \mathbf{C}\cdot(\mathbf{A}\times\mathbf{B})$$

(V-2) $$\mathbf{A}\cdot(\mathbf{B}\times\mathbf{C}) = (\mathbf{A}\times\mathbf{B})\cdot\mathbf{C} = -(\mathbf{B}\times\mathbf{A})\cdot\mathbf{C}$$

"BAC CAB" expansion

(V-3) $$\mathbf{A}\times(\mathbf{B}\times\mathbf{C}) = \mathbf{B}(\mathbf{A}\cdot\mathbf{C}) - \mathbf{C}(\mathbf{A}\cdot\mathbf{B})$$

"Do nothing" identity

(V-4) $$(\mathbf{A}\cdot\nabla)\mathbf{r} = -(\mathbf{A}\cdot\nabla)\mathbf{r}' = \mathbf{A}$$

Identities for the calculation of gradient

(V-5) $$\nabla(\varphi U) = \varphi\nabla U + U\nabla\varphi$$

(V-6) $$\nabla(\mathbf{A}\cdot\mathbf{B}) = (\mathbf{A}\cdot\nabla)\mathbf{B} + \mathbf{A}\times(\nabla\times\mathbf{B}) + (\mathbf{B}\cdot\nabla)\mathbf{A} + \mathbf{B}\times(\nabla\times\mathbf{A})$$

(V-7) $$\nabla\varphi(U_1\cdots U_n) = \sum_{i=1}^{n}\frac{\partial\varphi}{\partial U_i}\nabla U_i$$

Identities for the calculation of divergence

(V-8) $$\nabla\cdot(\varphi\mathbf{A}) = \varphi\nabla\cdot\mathbf{A} + \mathbf{A}\cdot\nabla\varphi$$

(V-9) $$\nabla\cdot(\mathbf{A}\times\mathbf{B}) = \mathbf{B}\cdot\nabla\times\mathbf{A} - \mathbf{A}\cdot\nabla\times\mathbf{B}$$

(V-10) $$\nabla\cdot\mathbf{A}(U_1\cdots U_n) = \sum_{i=1}^{n}\nabla U_i\cdot\frac{\partial\mathbf{A}}{\partial U_i}$$

Identities for the calculation of curl

(V-11) $$\nabla \times (\varphi \mathbf{A}) = \varphi \nabla \times \mathbf{A} + \nabla \varphi \times \mathbf{A}$$

(V-12) $$\nabla \times (\mathbf{A} \times \mathbf{B}) = (\mathbf{B} \cdot \nabla)\mathbf{A} + \mathbf{A}(\nabla \cdot \mathbf{B}) - (\mathbf{A} \cdot \nabla)\mathbf{B} - \mathbf{B}(\nabla \cdot \mathbf{A})$$

(V-13) $$\nabla \times \mathbf{A}(U_1 \cdots U_n) = \sum_{i=1}^{n} \nabla U_i \times \frac{\partial \mathbf{A}}{\partial U_i}$$

Repeated application of ∇

(V-14) $$\nabla \cdot (\nabla \times \mathbf{A}) = 0$$

(V-15) $$\nabla \times \nabla U = 0$$

(V-16) $$\nabla \times (\nabla \times \mathbf{A}) = \nabla(\nabla \cdot \mathbf{A}) - \nabla^2 \mathbf{A}$$

Identities for the calculation of line and surface integrals

(V-17) $$\oint \mathbf{A} \cdot d\mathbf{l} = \int \nabla \times \mathbf{A} \cdot d\mathbf{S}$$ (Stokes's theorem)

(V-18) $$\oint U d\mathbf{l} = \int d\mathbf{S} \times \nabla U$$

Identities for the calculation of surface and volume integrals

(V-19) $$\oint \mathbf{A} \cdot d\mathbf{S} = \int \nabla \cdot \mathbf{A}\, dV$$ (Gauss's theorem)

(V-20) $$\oint U d\mathbf{S} = \int \nabla U\, dV$$

(V-21) $$\oint \mathbf{A} \times d\mathbf{S} = -\int \nabla \times \mathbf{A}\, dV$$

(V-22) $$\oint (\mathbf{A} \cdot \mathbf{B}) d\mathbf{S} - \oint \mathbf{B}(\mathbf{A} \cdot d\mathbf{S}) - \oint \mathbf{A}(\mathbf{B} \cdot d\mathbf{S})$$
$$= \int [\mathbf{A} \times (\nabla \times \mathbf{B}) + \mathbf{B} \times (\nabla \times \mathbf{A}) - \mathbf{A}(\nabla \cdot \mathbf{B}) - \mathbf{B}(\nabla \cdot \mathbf{A})]\, dV$$

(V-23) $$\oint \mathbf{A}(\mathbf{B} \cdot d\mathbf{S}) = \int [(\nabla \cdot \mathbf{B})\mathbf{A} + (\mathbf{B} \cdot \nabla)\mathbf{A}]\, dV$$

Helmholtz's (Poisson's) theorem

(V-24) $$\mathbf{V} = -\frac{1}{4\pi}\int_{\text{All space}} \frac{\nabla'(\nabla'\cdot\mathbf{V}) - \nabla'\times(\nabla'\times\mathbf{V})}{r}\,dV'$$

Operations with ∇ *in Helmholtz's (Poisson's) integrals*

(V-25) $$\nabla'\frac{(\mathrm{X})}{r} = \frac{\nabla'(\mathrm{X})}{r} + \mathbf{r}_u\frac{(\mathrm{X})}{r^2}$$

(V-26) $$\nabla\frac{(\mathrm{X})}{r} = -\mathbf{r}_u\frac{(\mathrm{X})}{r^2}$$

(V-27) $$\frac{\nabla'(\mathrm{X})}{r} = \nabla\frac{(\mathrm{X})}{r} + \nabla'\frac{(\mathrm{X})}{r}$$

Retarded (causal) integrals

(V-28) $$\mathbf{V} = -\frac{1}{4\pi}\int_{\text{All space}} \frac{\left[\nabla'(\nabla'\cdot\mathbf{V}) - \nabla'\times(\nabla'\times\mathbf{V}) - \dfrac{1}{c^2}\dfrac{\partial^2\mathbf{V}}{\partial t^2}\right]}{r}\,dV'$$

(V-29) $$\mathbf{V} = -\frac{1}{4\pi}\int_{\text{All space}} \frac{\left[\nabla'^2\mathbf{V} - \dfrac{1}{c^2}\dfrac{\partial^2\mathbf{V}}{\partial t^2}\right]}{r}\,dV'$$

Operations with ∇ *in retarded (causal) integrals*

(V-30) $$\nabla'[\mathrm{X}] = [\nabla'\mathrm{X}] + \frac{\mathbf{r}_u}{c}\frac{\partial[\mathrm{X}]}{\partial t}$$

(V-31) $$\nabla[\mathrm{X}] = -\frac{\mathbf{r}_u}{c}\frac{\partial[\mathrm{X}]}{\partial t}$$

(V-32) $$[\nabla'\mathrm{X}] = \nabla[\mathrm{X}] + \nabla'[\mathrm{X}]$$

(V-33) $$\frac{[\nabla'\mathrm{X}]}{r} = \nabla\frac{[\mathrm{X}]}{r} + \nabla'\frac{[\mathrm{X}]}{r}$$

(V-34) $$\nabla\frac{[\mathrm{X}]}{r} = -\frac{\mathbf{r}_u[\mathrm{X}]}{r^2} - \frac{\mathbf{r}_u}{rc}\left[\frac{\partial\mathrm{X}}{\partial t}\right]$$

INDEX

Corrections to

ELECTROMAGNETIC RETARDATION AND THEORY OF RELATIVITY

by Oleg D. Jefimenko

P. 43, Line 3 from bottom: Change "squares" to "squares of the components"

P. 43, last line: Change z^2 to x^2.

P. 50, Fig. 3.2: Change θ to ϕ.

P. 68, Lines 6 and 9: Change (3-1.27) to (3-1.26).

P. 96, Line 6 from bottom: Change 1.27) to 1.26).

P. 98, Line 7: Change "puor" to "pour".

P. 108, Eq. (5-1.16): Change x to x'.

P. 125, Eq. (5-3.9): Change $L/2$ to L.

P. 125, Line 16: Change "differentiate" to "use Eqs. (2-4.8) and (5-2.12) and differentiate".

P. 130, Line 11: Change "1638" to "1632".

P. 137, Lines 15, 16: Change 4-5 and 4-5.1to 4-6 and 4-6.1.

P. 165, Lines 18, 21 and Eq. (7-5.1): Change $\mathbf{v}$ to $\mathbf{u}$.

P. 169, Eq. (7-5.24): Change $(1\text{-}vu/c^2)^2x$ to $(1\text{-}vu/c^2)x$.

P. 187, Last 4 lines: The explanation is inaccurate.

P. 191, Line 7 from bottom: Change $v = u_z$ to $v = u_x$.

P. 194, Eq. (8-4.7): Change $^{3/2}$ to $^{1/2}$.

P. 195, Line 3: Change $\mathbf{v})\mathbf{v} = Fu^2$ to $\mathbf{u})\mathbf{u} = \mathbf{F}u^2$.

P. 195, Eq. (8-4.8): Change $^{3/2}$ to $^{1/2}$ and $^{1/2}$ to $^{3/2}$.

P. 195, Line 15: Change 4.8) to 4.7).

P. 200, Line 18: Change "Eqs. (8-5.2) and (8-5.6)-(8-5.8)" to "Eqs. (8-5.10) and (8-5.14)-(8-5.16)".

P. 201, Line 7: Change "Eq. (8-5.3)" to "Eq. (8-5.11)".

P. 213, Line 3 from bottom: Change "line" to "long line".

P. 214, Eq. (9-2.7): Change $\gamma\mu_0\sigma\mathbf{k}$ to $\gamma\mu_0\sigma v\mathbf{k}$.

P. 214, Line 16: Change "line" to "long line".

P. 224, Line 17: Change "Eq. (4-5.6)" to "Eq. (4-6.6)".

P. 224, Line 20: Change "Eq. (9-4.19)" to "Eq. (9-4.18)".

P. 225, Line 2: Change "Eqs. (4-5.5) and (4-5.6)" to "Eqs. (4-6.5) and (4-6.6)".

P. 225, Lines 7 and 8: The statement is incorrect.

P. 227, Lines 7: The statement is incorrect.

P. 232, Last line: Change "Hardonic" to "Hadronic"